Nanoimprinting and Its Applications

Nanoimprinting and Its Applications

edited by

Akihiro Miyauchi

JENNY STANFORD
PUBLISHING

Published by

Jenny Stanford Publishing Pte. Ltd.
Level 34, Centennial Tower
3 Temasek Avenue
Singapore 039190

Email: editorial@jennystanford.com
Web: www.jennystanford.com

British Library Cataloguing-in-Publication Data
A catalogue record for this book is available from the British Library.

Nanoimprint and Its Applications

ISBN 978-981-4800-37-2 (Hardcover)
ISBN 978-0-429-03192-2 (eBook)

Contents

Preface

Nanoimprinting is a new technology for fabricating nanometer-scale patterns. Since it was proposed in 1995 by Prof. Chou, nanoimprinting has grown rapidly. Now machines, resins, and molds for nanoimprinting are commercially available worldwide. And application fields of nanoimprinting are expanding to not only electronics but also optics, biology, and energy fields because nanoimprinting is a simple and convenient method for nanofabrication.

In this book, the fundamentals of nanoimprinting are explained in terms of materials, processes, and machines, in Chapters 2, 3, and 4. The applications of nanoimprinting in optics, biology, energy, and electronics are described in Chapter 10. The fundamentals will be helpful for understanding nanoimprinting from the perspective of advanced undergraduate and graduate courses. Note that as many practical examples of nanoimprinting as possible are included in the book. These examples will be useful for engineers involved in research and development of various devices using nanostructures.

As the author, I would like to express my appreciation to the contributors, Prof. emeritus Matsui of University of Hyogo and Prof. Y. Hirai of Osaka Prefecture University, Drs. H. Hiroshima of the National Institute of Advanced Industrial Science and Technology, H. Sato of Canon Inc., and M. Okada of Asahi Kasei Corp. I would also like to thank the management staff of the Materials Innovation Center, Hitachi, Ltd. In particular, I have been unfailingly supported for more than 17 years by the following laboratory managers: Drs. S. Azuhata, Y. Fukunaga, H. Kodama, Y. Kozono, K. Kimura, H. Kodama, S. Yamada, T. Miwa, H. Akahoshi, and Y. Aono. I have also been continuously supported by the following business-department managers: Drs. H. Sakaguchi and S. Kashimura of Hitachi Cable, Ltd.; M. Kaji of Hitachi Chemical Co., Ltd.; K. Takahashi and Y. Kondo of Hitachi Industries, Ltd.; Dr. K. Souma and K. Sakaue of Hitachi Industrial Equipment Systems Co., Ltd.; R. Kawabata, H. Koshi, and S. Fukuzono of Hitachi High-Technologies Corporation; and S. Nishiuchi and M. Nemoto of Hitachi Ltd. I also wish to thank my coresearchers Profs. S. Matsui, Y. Hirai,

T. Kubota, M. Shimomura, and K. Tanaka, as well as Drs. H. Hiroshima, T. Uemura, S. Nomura, and M. Mita. Most of the data in this book were obtained through the research activities of Hitachi, Ltd. I would also like to thank the members of the nanoimprint research team Drs. K. Kuwabara, K. Ohashi, M. Ogino, M. Hasegawa, T. Ando, H. Yoshida, Y. Tada, S. Ishii, R. Washiya, C. Haginoya, Y. Takamori, and S. Suzuki, as well as T. Yamasaki, Moriwaki, T. Saito, H. Sonoda, and K. Konishi, of Hitachi, Ltd. And I acknowledge Stanford Chong of Jenny Stanford Publishing for giving me the opportunity to publish my research results on nanoimprinting. Finally, I especially thank my wife, Nagako, my son, Ryo, and my daughter, Noriko, for their assistance in many ways.

Akihiro Miyauchi
2019

Chapter 1

Introduction

Akihiro Miyauchi
Tokyo Medical and Dental University, 2-3-10 Kanda-Surugadai, Chiyodaku, Tokyo 101-0062, Japan
miyauchi.ibb@tmd.ac.jp

1.1 Background

Nanofabrication technology can be categorized into two kinds: bottom up and top down. Bottom-up technology includes self-organization and/or self-assembly. The most familiar examples of such technology are found in the human body, that is, the human body is generated from a fertilized egg and grows by self-organization.

As for top-down technology, a shape is created on the basis of an artificial design. A typical example of such a technology is a large-scale integration (LSI) circuit. LSI circuits are produced by a technology called lithography. In particular, photolithography is now widely used for fabricating electric devices and display panels. However, the cost of extreme ultraviolet (UV) lithography equipment reaches over 50 million US dollars. Although photolithography can achieve patterning on a scale of less than 10 nm, its fields of applicability are limited by its high equipment cost.

Nanoimprint and Its Applications
Edited by Akihiro Miyauchi

ISBN 978-981-4800-37-2 (Hardcover), 978-0-429-03192-2 (eBook)
www.jennystanford.com

Nanoimprinting is one of the top-down technologies, but as shown in Fig. 1.1, its fabrication cost is considerably lower than the costs of other lithography methods. It is, therefore, spreading into application fields in which nanoscale patterns enhance device characteristics.

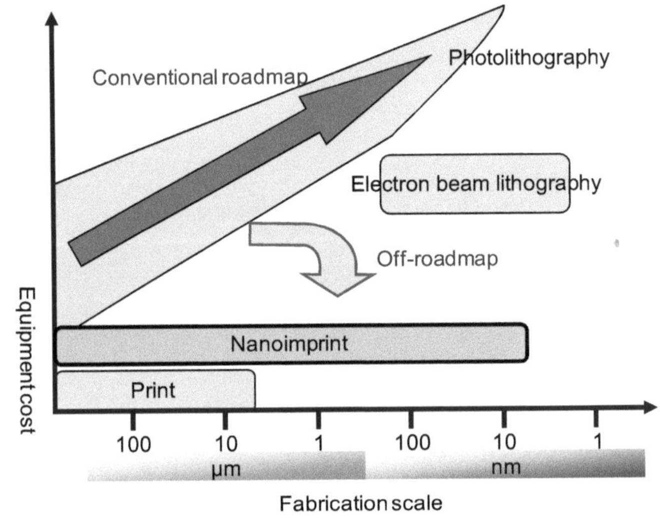

Figure 1.1 Positioning of nanoimprinting as a fabrication technology.

1.2 Principle of Nanoimprinting

Nanoimprinting is a method of press-molding on the nanoscale. Nanoscale patterns are fabricated by using a nanomold with fine concavity and convexity. The nanomold is pressed onto materials such as a resin layer on a substrate. As shown in Fig. 1.2, two main kinds of nanoimprinting are available, thermal nanoimprinting and photonanoimprinting.

Thermoplastic materials, such as polystyrene (PS) and acrylic resin, are used for thermal nanoimprinting. These polymer materials easily deform at high temperatures (typically 100°C–150°C). At a high temperature, the mold is pressed onto a thermoplastic layer and cooled. After cooling, the mold is removed (demolded) from the plastic layer. Thermal nanoimprinting is described in Chapter 3 in detail.

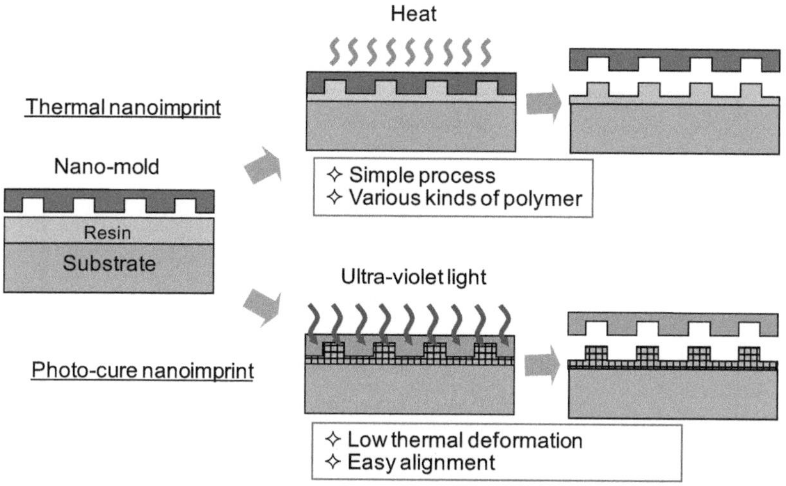

Figure 1.2 Principle of thermal nanoimprinting and photonanoimprinting.

In photonanoimprinting, photocurable resins are used as materials to be fabricated. Usually mixtures of monomers, a cross-linking agent, and a photoinitiator are used as photocurable resins, which are pressed and irradiated by UV light and cured. After the materials are cured, the mold is demolded from the materials. Photonanoimprinting is described in Chapter 4 in detail.

In both thermal nanoimprinting and photonanoimprinting, the principle is the same: the mold is pressed onto the materials to be fabricated and after the materials are cured, the mold is removed from the materials.

1.3 Applications

Nanoimprinting is expected to be applied in the early stages of manufacturing of electronic devices, such as storage media and LSI circuits [1–5]. As nanoimprinting progresses, researchers in various fields are showing a growing interest in nanofabrication.

As shown in Fig. 1.3, optical devices control visible light so that structures between several tens of nanometers and several micrometers become necessary to control visible light. The storage bits should be as small as possible to increase the memory capacity of the device. Gates and wires in LSI circuits are also becoming narrower and narrower over successive generations.

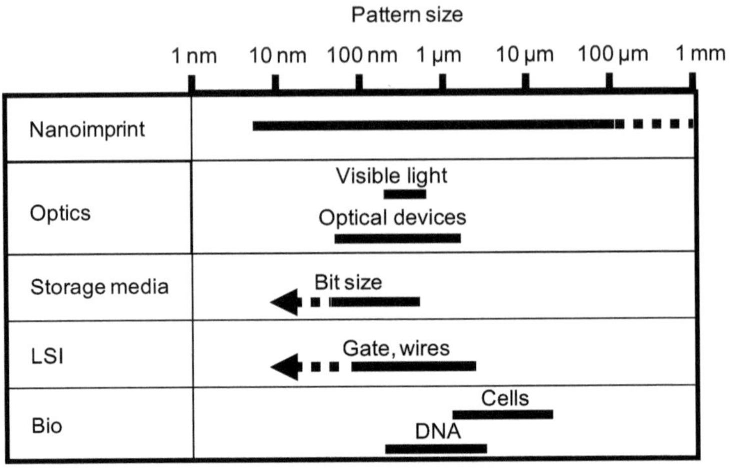

Figure 1.3 Nano- and microscale structures for various device fields.

In biology, cells and stretched DNAs are several microns to several dozen microns in size, so structures that are less than a micron in scale are attractive for manipulating these biological materials directly. Applications of nanoimprinting in the fields of optics, biology, energy, and electronics are described in Chapter 10.

References

1. S. Y. Chou, P. R. Krauss, and P. J. Renstrom (1995). Imprint of sub-25 nm vias and trenches in polymers, *Appl. Phys. Lett.*, **67**(21), 3114–3316.
2. S. Y. Chou, P. R. Krauss, and P. J. Renstrom (1996). Nanoimprint lithography, *J. Vac. Sci. Technol. B*, **14**(6), 1–5.
3. W. Wu, B. Cui, X. Y. Sun, W. Zhang, L. Zhunag, and S. Y. Chou (1998). Large-area high-density quantized magnetic disks fabricated using nanoimprint lithography, *J. Vac. Sci. Technol. B*, **16**(6), 3825–3829.
4. W. Wu, B. Cui, X. Y. Sun, W. Zhang, L. Zhunag, and S. Y. Chou (1998). Large area high density quantized magnetic disks fabricated using nanoimprint lithography, *J. Vac. Sci. Technol. B*, **16**(6), 3825–3829.
5. M. Colburn, S. Johnson, S. Damle, T. Bailey, B. Choi, M. Wedlake, T. Michaelson, S. V. Sreenivasan, J. Ekerdt, and C. G. Willson (1999). Step and flash imprint lithography: a new approach to high-resolution patterning, *Proc. SPIE*, **3676**, 379–389.

Chapter 2

Template Technology

Akihiro Miyauchi
Tokyo Medical and Dental University, 2-3-10 Kanda-Surugadai, Chiyodaku, Tokyo 101-0062, Japan
miyauchi.ibb@tmd.ac.jp

Many kinds of materials are used as nanoimprinting molds, which are obtained by replicating original templates. The template is fabricated by photolithography, anodic oxidation, self-organization, and so on. Multiple molds are replicated from one template.

The molds are usually pretreated to prevent the imprinted materials (such as polymers and resists) from sticking to the mold. In this chapter, methods for fabricating templates are introduced and typical antisticking treatments for molds are explained. The methods for replicating molds from templates for thermal nanoimprinting and photonanoimprinting are described in Chapters 3 and 4, respectively.

Nanoimprint and Its Applications
Edited by Akihiro Miyauchi

ISBN 978-981-4800-37-2 (Hardcover), 978-0-429-03192-2 (eBook)
www.jennystanford.com

2.1 Template Fabrication

A template is fabricated in several ways. The most suitable method depends on pattern layout, pattern size, and template materials. As methods for fabricating the template photolithography, anodic oxidation, and self-organization are explained in the following sections.

2.1.1 Photolithography

Several kinds of template materials are available. The most standard material is silicon because it can be fabricated by a conventional large-scale integration (LSI) process. As shown in Fig. 2.1, UV light is partially transmit to the chromium/quartz plate and focused on the resist layer on a silicon wafer. Fine patterns are formed on the silicon wafer by the dry etching process. The minimum pattern size F is given as

$$F = k\frac{\lambda}{N_{\mathrm{A}}}, \tag{2.1}$$

where k is a process-oriented factor (generally 0.5–0.6), λ is the wavelength of the UV light, and N_{A} is the aperture ratio. To form a fine pattern, λ must be short. Light sources and their wavelengths over each generation are listed in Table 2.1. Advanced integrated circuits (ICs) created by LSI, known as LSI circuits, are produced by photolithography using an ArF excimer laser, and photolithography equipment using EUV (extreme UV) light is now under development.

Table 2.1 Light source and wavelength for photolithography

Light source	High-pressure mercury vapor g-line	High-pressure mercury vapor i-line	KrF excimer laser	ArF excimer laser	EUV (under development)
Wavelength (nm)	436	365	248	193	13.5

An example of a silicon template is shown in Fig. 2.2. The diameter of the template is limited by the wafer size used in LSI circuit production. The die size is several dozen square millimeters, and dies are periodically laid out on the silicon wafer. Tiny dots and lines are formed on the silicon template, so the surface of the template appears in the colors of a rainbow due to the optical-grating phenomenon. An example of the surface structure of the silicon template is shown in Fig. 2.2b. In this example, holes 180 nm in diameter are formed on the surface. These holes are fabricated by photolithography and dry etching.

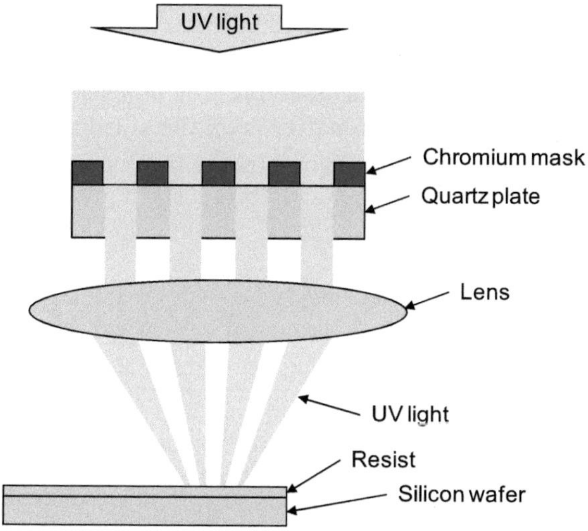

Figure 2.1 Schematic diagram of photolithography.

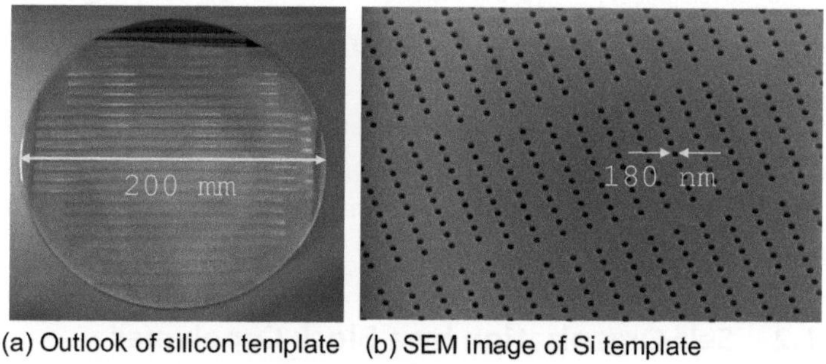

Figure 2.2 Outlook of silicon template and its SEM image. SEM, scanning electron microscopy.

The dry-etching process is shown schematically in Fig. 2.3. A chemically treated clean silicon wafer is prepared, and a resist layer is formed on the wafer by spin coating (step b). The resist layer is partially cured by photocuring using a mask (step c). After the cured resist stabilizes, it is rinsed and the noncured area is removed by organic solutions (step d). When the UV-irradiated areas remain, the resist is called "negative resist." On the other hand, if the UV-irradiated areas are dissolved during the development process, the resist is called "positive resist." After the resist is developed, the silicon wafer is etched by the plasma of CF_4 or C_2F_6 (step e). The areas of the silicon surface that are covered by the resist are not etched. After the resist is stripped off, tiny holes are formed on the silicon surface (step f). The patterns on the silicon surface, such as lines and corners, can be designed by the mask patterns used in the photocuring process.

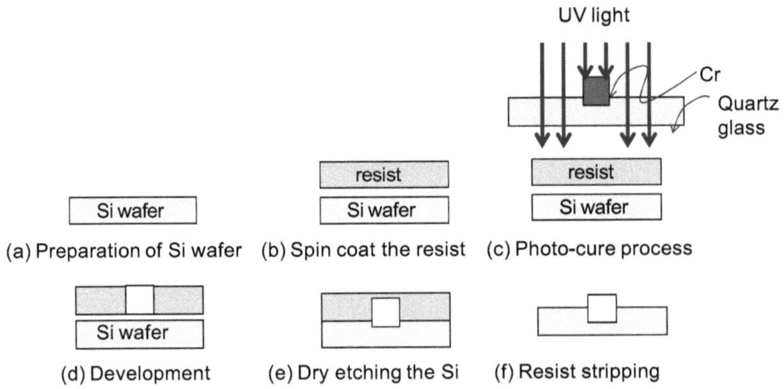

Figure 2.3 Dry process sequence of hole fabrication on a silicon wafer.

As shown in Fig. 2.4, a transparent material, such as glass, can also be fabricated by this photolithography method. A transparent template is suitable for photonanoimprinting, which is described in detail in Chapter 4.

2.1.2 Self-Organization by a Block Copolymer

Fabricating the template by photolithography is convenient because it is an established fabrication process allowing a high degree of design freedom. However, the fabrication area by photolithography is limited by the exposure area of the lithography equipment. Moreover, the maximum diameter of the silicon wafer is 300 mm.

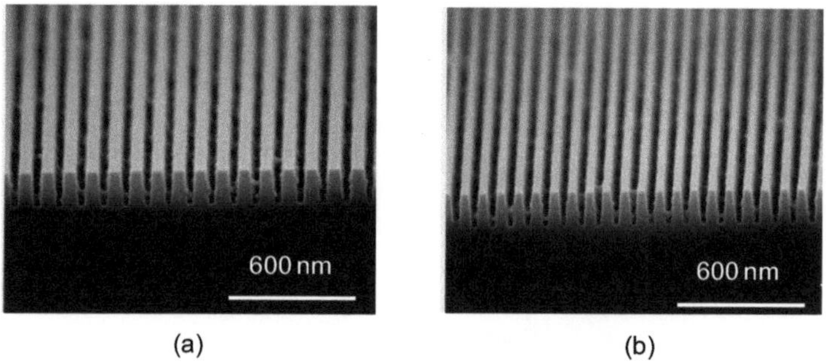

Figure 2.4 SEM image of quartz template for photonanoimprinting. Line pitch is (a) 100 nm and (b) 80 nm. SEM, scanning electron microscopy.

To form a fine pattern over a large area, self-organization is a well-known method as a bottom-up technology. As shown in Fig. 2.5a, a block copolymer (which consists of two different natural polymers) is used as a source material. Polymer chains A and B interact differently with a solvent, and a so-called χ parameter defines the segregation of these chains in a solvent. The initial state of the block copolymers, namely randomly distributed, is shown in Fig. 2.5b. On annealing, as shown in Fig. 2.6, the block copolymers make self-organized structures. The driving force of the shape formation is the difference in the affinity of each polymer with the solvent and with the other polymer. The cylindrical and lamellar structures are used in nanoimprinting templates.

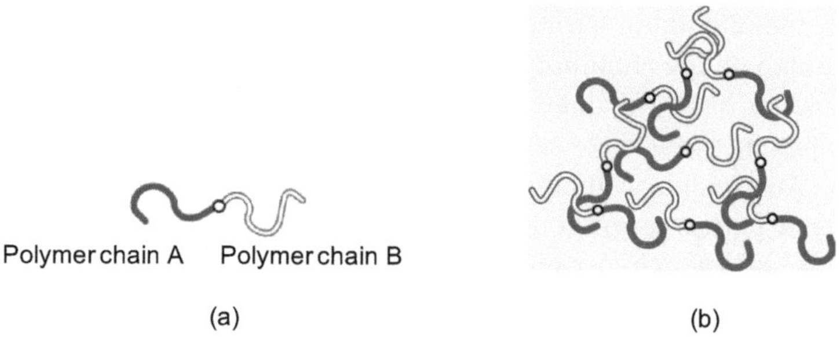

Figure 2.5 Schematics of (a) a single block copolymer and (b) block copolymers in random state.

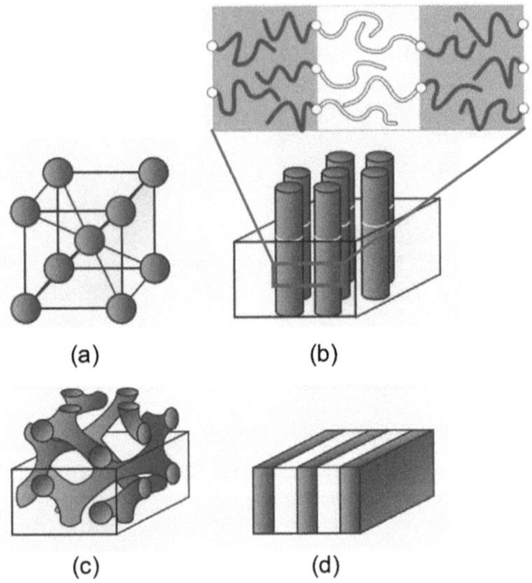

Figure 2.6 Typical structures of self-organization of block copolymers. (a) Sphere, (b) cylinder, (c) bicontinuous, and (d) lamellar.

2.1.3 Self-Organization by Anodic Oxidation

Anodic oxidation is a well-known process for forming a periodic hole pattern. A schematic diagram of anodic-oxidation equipment is shown in Fig. 2.7. An aluminum sample and a carbon electrode are dipped in an electrolytic solution, such as oxalic acid or diluted sulfuric acid. A voltage is applied between the aluminum sample and the carbon electrode; as a result, an anodic oxidation film is formed on the aluminum surface. Tiny holes with a high aspect ratio are formed on the aluminum surface by this process. The hole layout is random, but the diameter and pitch of the holes can be controlled by adjusting the acid concentration and the voltage, respectively.

An example of anodic oxidation of aluminum is shown in Fig. 2.8 [1]. A pattern of periodic holes was obtained (image a), and each hole has a very high aspect ratio (image b).

A template made by anodic oxidation of aluminum is mechanically durable, so the aluminum template itself can be used as an imprinting mold for mass production. An aluminum-roll template has been used in photonanoimprinting, but reduced lifetime due to hole clogging by the photocurable resin is a problem. Release treatment is thus very important in cost reduction using an anodic-oxidation template.

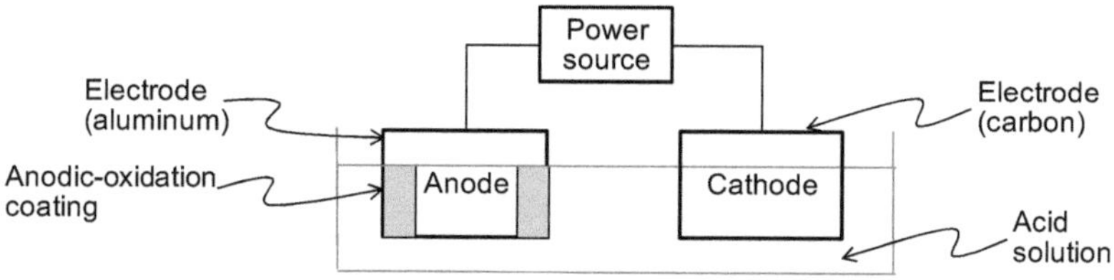

Figure 2.7 Schematic diagram of equipment for anodic oxidation.

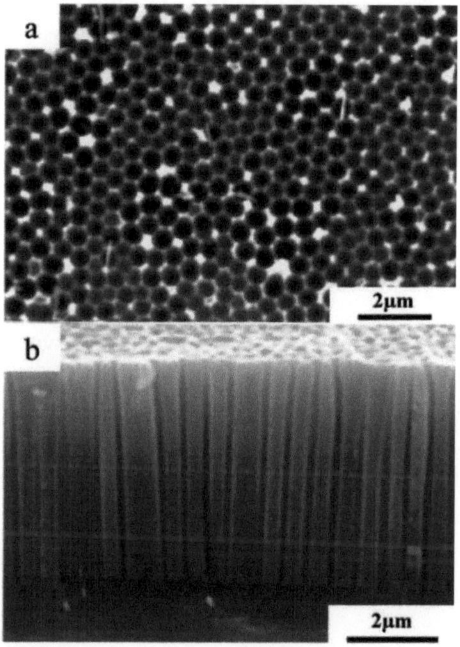

Figure 2.8 SEM images of anodic oxidation film. (a) Plane and (b) cross section. SEM, scanning electron microscopy. Republished with permission of The Electrochemical Society, from Ref. [1] copyright (2004); permission conveyed through Copyright Clearance Center, Inc.

2.1.4 Interference Exposure

Interference exposure is also suitable for obtaining a large-area mold. It uses interference of two coherent lights. A schematic diagram of interference exposure is shown in Fig. 2.9. Interference exposure generally uses two laser lights, and interference on the resist surface causes strong and weak exposure lines periodically. These periodical exposure lines can be used to form a line-and-space or a dot pattern in a large area. This method uses interference of two coherent lights, so the periodicity is limited by the wavelength of the laser.

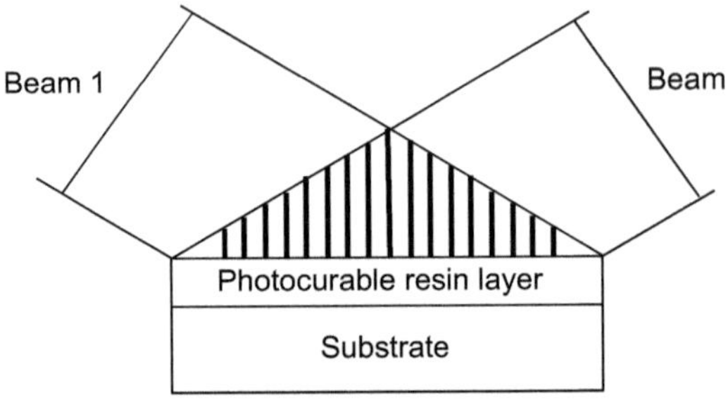

Figure 2.9 Schematic diagram of interference exposure.

2.2 Release Process

Nanoimprinting involves a contact process between a mold and an imprinted material. Integrity of the demolding process is, therefore, a critical issue for defect generation. In nanoimprinting, molds are processed by a release treatment before mold pressing. The release treatment of the mold and its degradation and a solution to recover the release property are described in this section.

2.2.1 Release Layer

One of the most important processes in nanoimprinting is demolding, that is, separating the mold and the molded resin. In general, separation of the two materials depends on the surface energies of the materials. In the case of a liquid on a solid substrate, as shown in Fig. 2.10, the state of the liquid becomes either wetting or de-wetting. The preferable state depends on the total energy of the system. The contact angle between the solid and the liquid, θ, is described by Young's formula as

$$\gamma_S = \gamma_{SL} + \gamma_L \cos\theta, \tag{2.2}$$

where γ_S, γ_{SL}, and γ_L are the surface energies of the solid, the liquid, and the interface between the solid and the liquid, respectively. Equation 2.1 indicates that the state of the liquid is de-wetting, that is, "easy" separation mode, when γ_S is small. Therefore, to separate the mold from the resin layer, the surface energy of the mold surface should be low. In nanoimprinting, the mold surface is generally covered by a low-surface-energy layer, that is, a release layer.

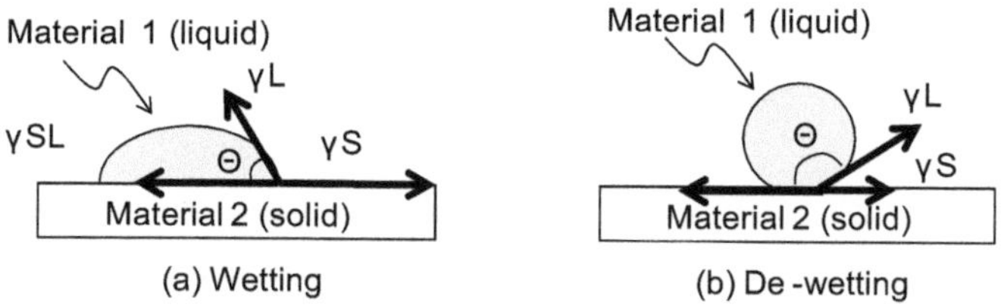

Figure 2.10 Schematics of two states of liquid on a solid surface. (a) Wetting and (b) de-wetting.

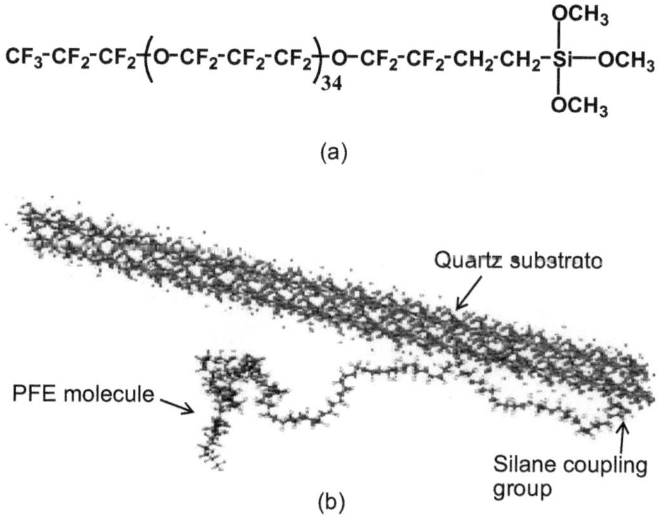

Figure 2.11 (a) Molecular formula and (b) structure of the typical release molecule, polyfluoroether (PFE), with a silane coupling group. The thickness is about 2 nm. The molecular structure is calculated by first-principles calculation.

The molecular formula and structure of a typical release molecule, for example, polyfluoroether (PFE) with a silane coupling group, are shown in Fig. 2.11. The main structure of this material is a CF_2 chain, and CF_2 chains cover the mold surface (i.e., a quartz substrate, as shown in Fig. 2.11) and reduce the mold surface energy. The silane coupling group reacts with the hydroxyl group of a mold surface. After dehydration synthesis, PFE molecules stick strongly on the mold surface by covalent bonds. Several other release molecules are reported [2]. The molecules with shorter CF_2 chains have the advantage of conformal coverage, and Cl-coupling groups are more strongly connected to metal surfaces. Release materials should be selected properly on the basis of the mold materials, patterns, and imprinted materials.

2.2.2 Degradation of the Release Layer

The release layer is exposed to severe conditions during nanoimprinting: thermal stress in thermal nanoimprinting, radical attack in photonanoimprinting, and mechanical stress during the release process. Atomic force microscopy (AFM) images of the 1st and the 150th nanoimprinting are shown in Fig. 2.12. The imprinted pattern is a pillar structure. The 150th imprint lacks some pillars, namely defects. These defects are generated by the resin's plugging up the holes of the mold due to degradation of the release layer.

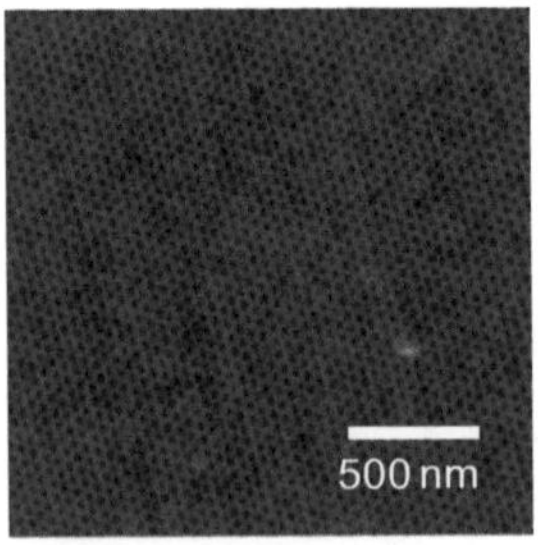

(a) 1st nanoimprint

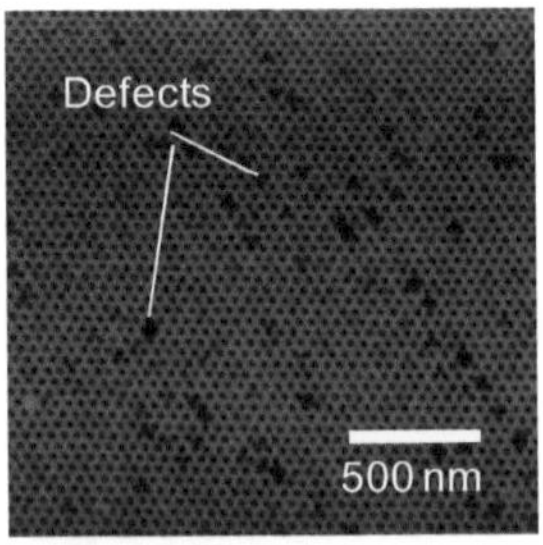

(b) 150th nanoimprint

Figure 2.12 Defect generation during the nanoimprinting cycle. AFM images of nanoimprinting sample. AFM, atomic force microscopy.

The release layer is extremely thin, so its thickness is evaluated by X-ray reflectivity (XRR). The principle of XRR is shown in Fig. 2.13. An incident X-ray beam is reflected at the surface of the release layer, and the interface between the release layer and the substrate. As shown in Fig. 2.14, the reflected X-ray beams interfere and make a so-called beat in the X-ray-diffractively curve, where **q** is the scattering vector, θ is the incident angle, and λ is the wavelength of the X-ray beam. The thickness and electron density of the thin layer can be calculated from the period and amplitude of the beat, respectively. In this calculation, the thickness and the density of the PFE release layer were 2.2 nm and 1.7 g/cm^3, respectively.

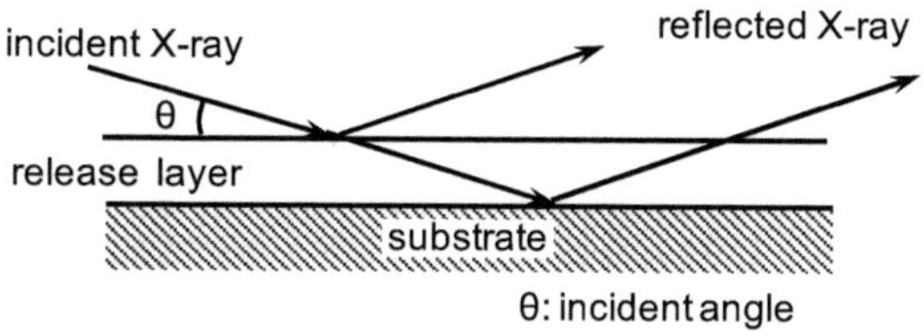

Figure 2.13 Principle of XRR. XRR, X-ray reflectivity.

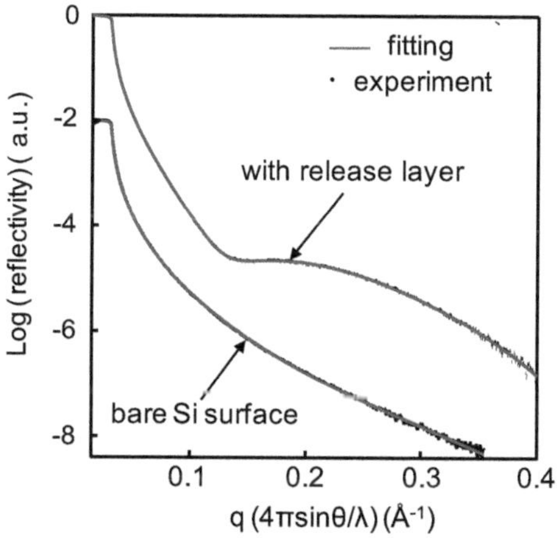

Figure 2.14 Observed XRR. The beat is observed in the sample with a release layer. XRR, X-ray reflectivity.

The XRR and calculated thickness of the release layer are plotted in Fig. 2.15 as a function of the number of nanoimprinting cycles. Clearly, as the number of nanoimprinting cycles increases, the beats in the X-ray diffractively become smaller, as shown in Fig. 2.15a. In other words, the release layer becomes thinner with an increasing number of nanoimprinting cycles. The calculated thickness of the release layer becomes about half at the 150th nanoimprinting cycle. This decrease in the release layer thickness generates the sticking of the resist to the mold, and the defects appear as shown in Fig. 2.12. There are several ways to prevent the degradation of the release layer, such as modifying the molecular structure of the release layer, choosing the coupling group optimized in regard to the chemical states of the substrate surface, and incorporating a release component in the resin material. The durability of the release layer relates to the defect generation and is a key factor in mass production with nanoimprinting.

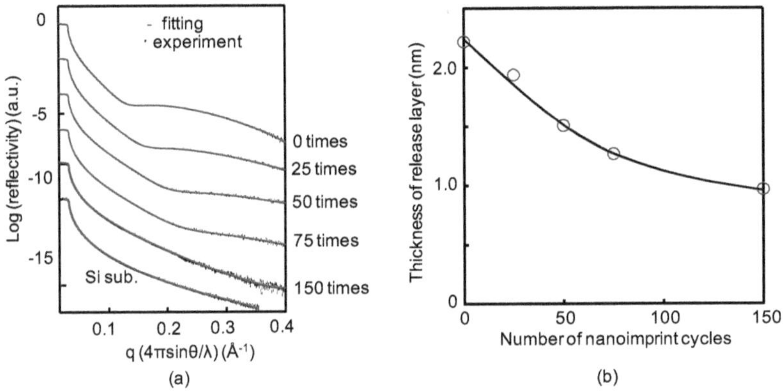

Figure 2.15 Measured thickness of release layer by XRR. (a) XRR and (b) obtained release layer thickness as a function of nanoimprint cycles. XRR, X-ray reflectivity.

2.2.3 Recovery of the Release Layer

As described in the previous section, the durability of the release layer is a key factor in high-quality nanoimprint. In this section, one of the methods to recover the release layer is introduced.

A reduction in the release layer thickness during thermal nanoimprinting is shown in Fig. 2.16. As for thermal nanoimprinting, the details are described in Chapter 3. In this process, the silicon wafer with a release layer is pressed onto a polystyrene (PS) film. As shown in the figure, the release layer thickness monotonically decreases with the number of nanoimprinting cycles. Atomic fractions of fluorine atoms on the imprinted PS films are shown in Fig. 2.17. New PS films are inserted in every nanoimprinting process in the experiments. The amount of fluorine atoms is measured by

X-ray photoelectron spectroscopy (XPS). The highest amount of fluorine atoms is observed at the first nanoimprinting cycle, after which the amount of fluorine atoms on the PS films gradually decreases. Comparing the data in Fig. 2.16 with that in Fig. 2.17 reveals that the release layer on the mold transfers to the surface of the PS film during the nanoimprinting. Therefore, keeping enough thickness of the release layer on the mold is one way to maintain the release property.

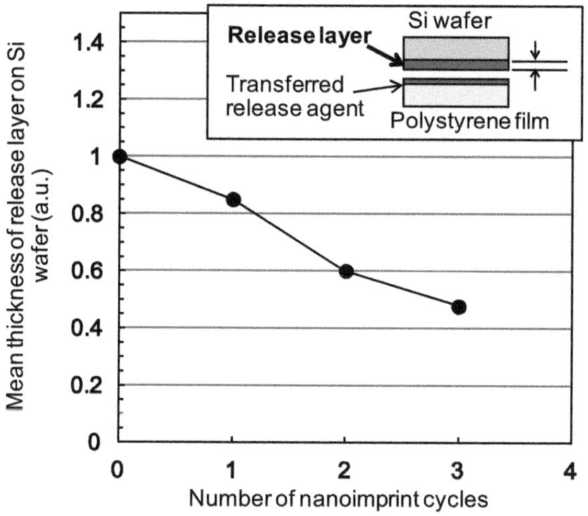

Figure 2.16 Reduction of release layer during thermal nanoimprinting. The normalized thickness of the release layer on a Si mold is evaluated by XRR. XRR, X-ray reflectivity.

A simple method to recover the release layer is introduced as follows [3]. The experimental setup for release layer recovery is shown in Fig. 2.18. The nanoimprinting process of the sheet-nanoimprinting system is described in Chapter 3. The mold used for sheet-nanoimprinting is belt shaped and rotates. The film is transferred from the left to the right in the figure. The mold and the film are pressed together by the press roll. The belt mold is pretreated by the previously described release treatment before setting. The unique point of this system is having a unit to supply the release agent. This unit sprays the release agent onto the belt-mold surface during a nanoimprinting cycle. The release molecules described in Section 2.2.1 have a reaction group, such as a silane coupling group. However, the release agent without the coupling group is used in this experiment because there is not enough time for dehydration synthesis. According to the results concerning the reduction in the release layer thickness shown in Figs. 2.16 and 2.17, the release layer thickness should be recovered every three nanoimprinting cycles.

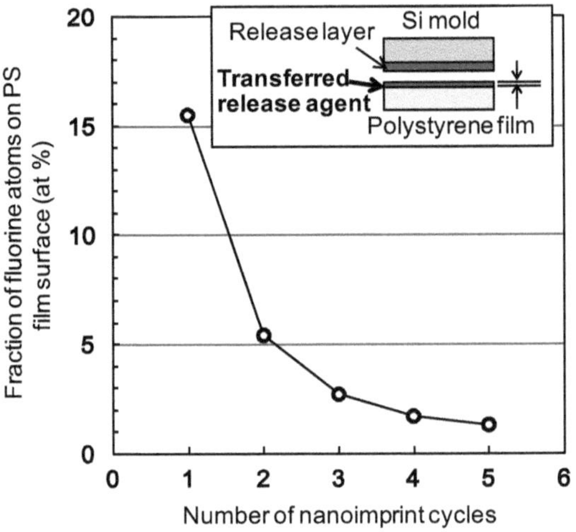

Figure 2.17 Atomic fraction of fluorine atoms on the PS film surface as a function of nanoimprint cycles. The atomic fraction of fluorine was measured by XPS. PS, polystyrene; XPS, X-ray photoelectron spectroscopy.

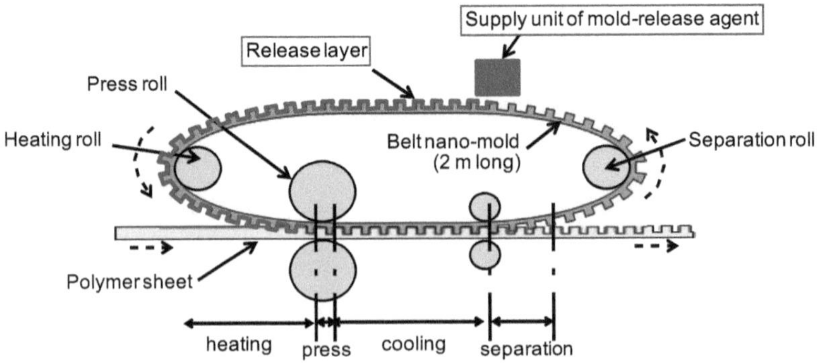

Figure 2.18 Experimental setup of release layer recovery.

The atomic fraction of fluorine atoms on the imprinted PS film surface is plotted as a function of the nanoimprinting cycle in Fig. 2.19. The release agent was sprayed every three nanoimprinting cycles. The fluorine fraction on the imprinted PS film decreases during the nanoimprinting cycles, but it is increased by supplying the release agent. The amount of transferred fluorine is enough compared to the initial state. Therefore, by repeating the spray process, it is possible to maintain the release property. The fidelity of the pillars of 200 m continuous nanoimprinting is shown in Fig. 2.20.

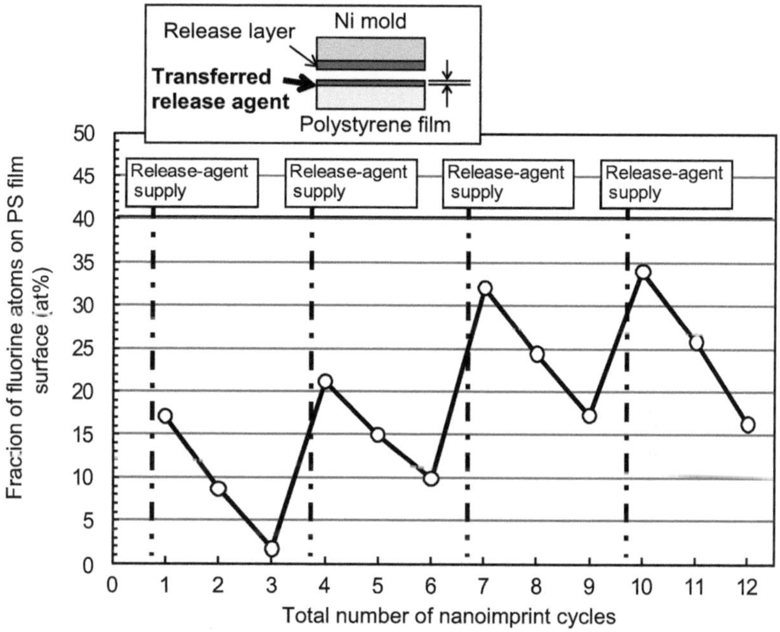

Figure 2.19 Fluorine fraction of imprinted PS film surface. The recovery of the release layer was examined every three imprint cycles.

The average pillar height is 1.05 μm, with a standard deviation, σ, of 0.03 μm. Continuous nanoimprinting to 1.8 km printing was confirmed by using the unit to supply the release agent. And the release layer recovery method described here is expected to maintain a high fidelity of the continuous nanoimprinted film.

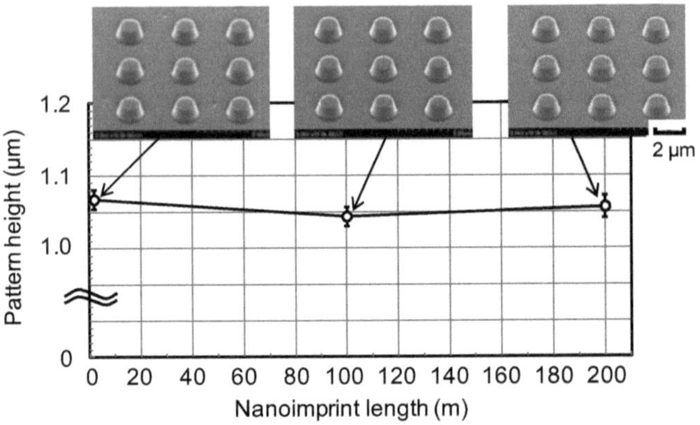

Figure 2.20 Imprint fidelity using a spray unit in continuous nanoimprinting.

References

1. S. Ono, M. Saito, M. Ishiguro, and H. Asoh (2004). Controlling factor of self-ordering of anodic porous alumina, *J. Electrochem. Soc.*, **151**, B473–B478.
2. M. Nakagawa (2015). *Nanoimprint Technology*, eds. S. Matsui, and Y. Hirai, Chapter 7: De-molding technology (The Institute of Electronics, Information and Communication Engineers, Tokyo), pp. 134–149 (in Japanese).
3. M. Ogino, M. Hasegawa, A. Miyauchi, K. Sakaue, and S. Nagai (2013). Fabrication of 200-nm dot pattern on 15-m-long polymer sheet using sheet nanoimprint method, *Jpn. J. Appl. Phys.*, **52**(3), 035201.1–035201.4.

Chapter 3

Thermal Nanoimprinting

Akihiro Miyauchi
Tokyo Medical and Dental University, 2-3-10 Kanda-Surugadai, Chiyodaku, Tokyo 101-0062, Japan
miyauchi.ibb@tmd.ac.jp

Thermal nanoimprinting uses thermoplastics as molding materials. Since various kinds of thermoplastic materials are available, various applications using thermal nanoimprinting are expected. In this chapter, the thermal-nanoimprinting process is explained in terms of materials, molds, and equipment. An actual application of thermal nanoimprinting is also introduced.

3.1 Process

The basic process sequence of thermal nanoimprinting is shown in Fig. 3.1. Thermoplastic resins are generally used as molding materials. The thermoplastic resin is heated to the nanoimprinting temperature, which is higher than the glass transition temperature of the resin. The resin is then pressed by a mold and cooled. When the mold temperature falls below the glass transition temperature, the mold is removed from the resin.

Nanoimprint and Its Applications
Edited by Akihiro Miyauchi

ISBN 978-981-4800-37-2 (Hardcover), 978-0-429-03192-2 (eBook)
www.jennystanford.com

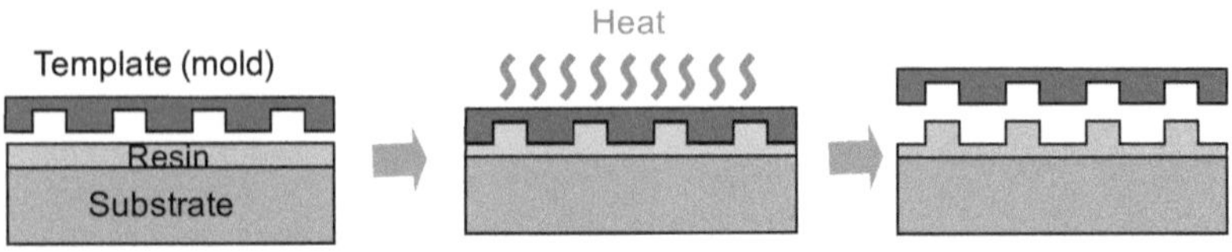

Figure 3.1 Process sequence of thermal nanoimprinting.

A typical elastic modulus *G* of a thermoplastic resin is plotted as a function of temperature in Fig. 3.2. Clearly, as the temperature increases, the elastic modulus of the thermoplastic resin falls gradually and then decreases drastically at a certain temperature, called the glass transition temperature. Around the glass transition temperature, the polymer chains have a high degree of freedom and start to move. The mold is pressed onto the resin at a temperature above the glass transition temperature and cooled after the appropriate press time. When the temperature drops below the glass transition temperature, the mold can be removed from the resin because the molecules in the resin become stable. The mold is usually pretreated by a "release treatment" for easy demolding (as described in Section 2.2).

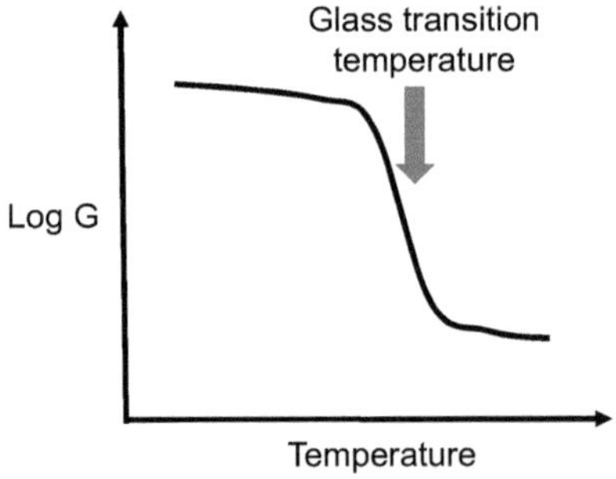

Figure 3.2 General behavior of shear modulus of thermoplastic materials.

3.2 Resin Materials

Some thermoplastic resins are listed in Table 3.1. The typical ones are polycarbonate (PC), polymethylmethacrylate (PMMA), and polyethylene terephthalate (PET). These materials are widely used in various industrial fields; therefore, thermal nanoimprinting is expected to be applied in a wide range of application fields.

Table 3.1 Characteristics of typical thermoplastic materials

Item	unit	PC	PMMA	PET	TOPAS®	ZEONOR®	ARTON®
Specific weight	g/cc	1.20	1.19	1.25	1.02	1.01	1.08
Transmittance	%	92	93	90	91	92	92
Refraction index	-	1.590	1.492	1.630	1.530	1.530	1.510
Water absorption coefficient	%	0.20	0.30	0.14	<0.01	<0.01	<0.01
Saturated water absorption	%	0.40	2.00	-	0.00	0.00	0.40
Glass transition temperature Tg	degree C	150	105	142	140	136	171
Linear expansion coefficient	/degree	7.0E-5	6.0E-5	7.0E-5	6.0E-5	7.0E-5	6.1E-5
Bending elastic modulus	kgf/cm^2	24,000	31,000	25,000	32,000	22,000	30,000
Bending strength	kgf/cm^2	930	1,050	1,500	-	940	1,250
Breaking elongation	%	90	5	23	4	20	15

The refractive indices of air and glass, which are used widely in optical devices, are 1.0 and 1.4–2.1, respectively. The refractive index of the resin is, therefore, an important factor in the optical design of devices. In addition, the water absorption coefficient is mainly related to the reliability of devices. And, as already described, the glass transition temperature is an important factor in thermal nanoimprinting. The linear expansion coefficient of the resin is also an important factor in the molding and demolding process. Differing linear expansion coefficients of the mold and resin cause thermal stress at the interface between them. Moreover, the bending elastic modulus and the bending strength are related to the mechanical properties of the resin. During the mold and resin separation process, several kinds of mechanical-stress-related phenomena, such as mold bending and sticking, can occur. Accordingly, resins for fine nanoimprinting must be mechanically tough. In particular, resins must stretch to their limit, namely breaking elongation, which is a property of bulk resins. In nanoimprinting, elongation phenomena occur in a constrained tiny space, so the elongation phenomena become unusual [4, 5].

The most important characteristic of thermoplastic resins for nanoimprinting is the glass transition temperature, which can be measured by a torque rheometer. In this measurement system, a thermoplastic sample is set between two flat plates, one of which is rotated by a small angle. The plate temperature is controllable. Elastic properties (namely storage elastic modulus, loss elastic modulus, relaxation time, and viscosity coefficient) are obtained as a function of temperature from amplitude ratio and phase delay.

3.3 Molds

In thermal nanoimprinting, a mold is heated up and pressed onto a resin. As described in Chapter 2, the template is usually made of silicon because silicon is commonly used in photolithography. In thermal nanoimprinting, however, silicon is unsuitable for pressing because a silicon mold is a crystal, so tiny solid particles in the press area cause the silicon to fracture. Therefore, a nickel mold is standard in thermal nanoimprint because nickel has fracture toughness and nickel plating is a well-established technology in injection molding, metal coating, and so on. Although not discussed in detail here, a carbon mold is used for glass nanoimprinting because of its thermal durability and good release property.

Nickel molds are obtained by electroplating. It is not preferable to plate nickel directly on a silicon template because a silicon template is easily damaged during the release process of nickel and silicon. Therefore, as shown in Fig. 3.3, a polymer replica is used as a buffer replica for nickel plating. A lot of polymer replicas can be obtained from one master template because a polymer replica is soft and seldom destroys the tiny structures on the silicon template. The polymer replicas are obtained by thermal nanoimprinting.

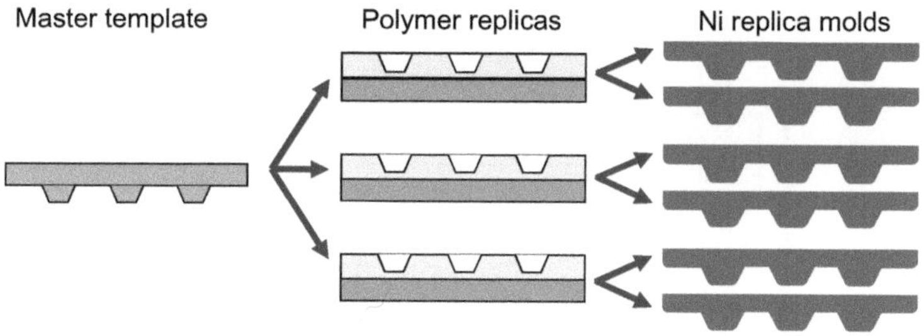

Figure 3.3 Schematic diagram of template replication.

Nickel replica molds are obtained by electroplating. A schematic diagram of nickel electroplating is shown in Fig. 3.4. A polymer replica is set as a cathode electrode. A polymer is an insulator, so metals like platinum are predeposited on the surface of the polymer replica by sputtering. A nickel electrode is immersed in the electroplating solution. Nickel ions are dissolved by applying a DC voltage. The dissolved nickel ions segregate on the surface of the polymer replica. The spatial resolution of nickel electroplating is quite high, so nanoscale replication is possible by nickel-ion segregation. After the nickel plating, the nickel foil is removed from the polymer replica and a nickel replica mold is obtained.

The appearances of the silicon template, the polymer replica, and the nickel replica mold are shown in Fig. 3.5. The silicon template is the size of a 200 mm wafer. A thermosetting polymer was used as the polymer replica material in this case because it is relatively stable in acid solutions. Scanning electron microscopy (SEM) images of each surface are shown in Fig. 3.6. Holes 180 nm in diameter are formed on the silicon template. By fine replication with a polymer, pillars with a diameter of 180 nm and a height of 400 nm are formed on the replica surface. The depth of the holes on the silicon template is designed to be 400 nm, so the height of the pillars is also 400 nm. After nickel electroplating, the same surface morphology as that of the silicon template was obtained. The polymer replica can be obtained from the silicon template many times; therefore, once an expensive and complex silicon template is fabricated, the replication process reduces the cost of the mold drastically.

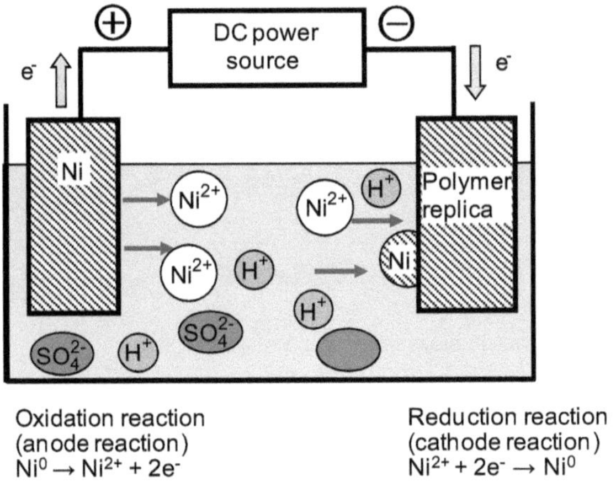

Figure 3.4 Schematic diagram of Ni electroplating.

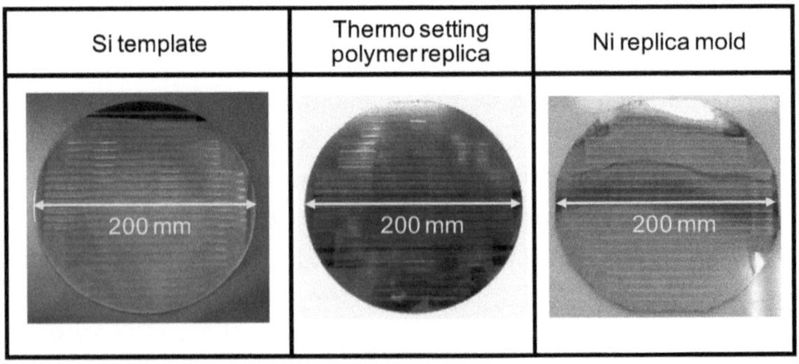

Figure 3.5 Appearance of a Si template, polymer replica, and Ni replica mold.

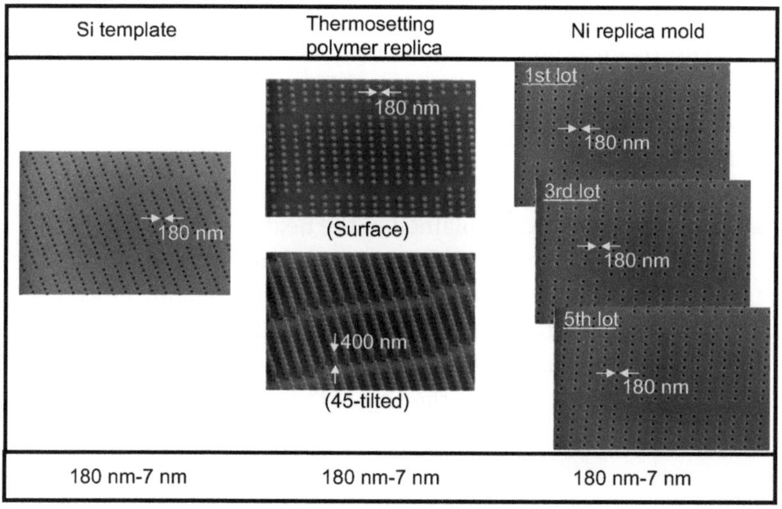

Figure 3.6 SEM image of each surface of Fig. 3.5. SEM, scanning electron microscopy.

3.4 Equipment

There are mainly two types of nanoimprinting equipment for both thermal nanoimprinting and photonanoimprinting, the parallel press and the roll press. Images of parallel-press nanoimprinting are shown in Fig. 3.7. In this case, the mold is pressed parallel to the substrate; consequently, the mold patterns are usually transferred at one time, as shown in Fig. 3.7a. When relatively large imprints are necessary, because the mold size is limited, the mold is repositioned and sets of imprints are pressed one by one to obtain a large imprint area, as shown in Fig. 3.7b. In the case of Fig. 3.7b, the positioning of the mold on a prepatterned area—called alignment—is an important factor.

The other type of nanoimprinting, using the roll press, is shown in Fig. 3.8. Three types of mold are used: a patterned-roll mold (Figs. 3.8a and 3.8b), a rotating-belt mold (Figs. 3.8c and 3.8d), and a fixed-foil mold (Fig. 3.8e).

In Figs. 3.8a and 3.8b, a roll mold is used. The roll mold can be manufactured by two methods: (i) sticking a thin sheet mold around a roll [1] and (ii) forming patterns directly on the roll surface by electron beam lithography (EBL) [2] or anodic oxidation. Method (i) uses a patterned round sheet, so a boundary appears at the sheet edge. Method (ii) has no boundary on the roll surface, so continuous patterning is possible.

In Figs. 3.8c and 3.8d, a belt-shaped mold is necessary. An example of this type is explained in the next section.

In Fig. 3.8e, the mold is a conventional foil type but the pressing is done by a roll press. When a roll press is used, air does not get trapped between the mold and the substrate because the mold is in contact with the substrate linearly, not as a plane. However, the printable area is limited by the size of the foil mold.

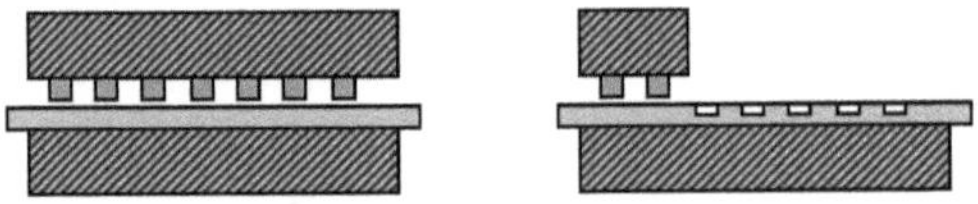

(a) (b)

Figure 3.7 Press image of parallel press–type nanoimprint. (a) The pattern is formed at one time, but the patterned area is limited by mold size. (b) A large area can be imprinted by the step-and-repeat action of the mold, but alignment is necessary to control the mold position.

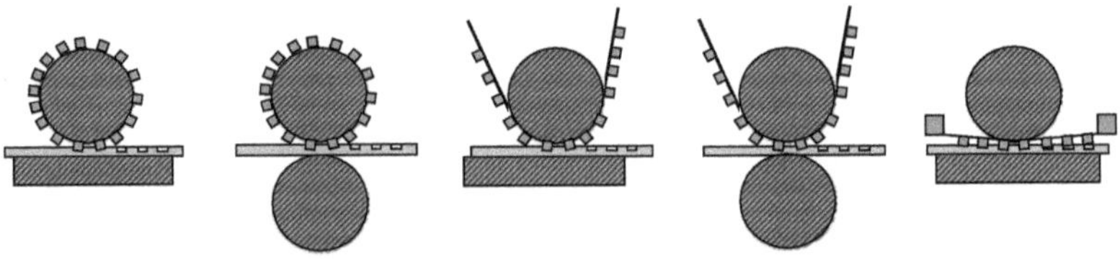

(a) (b) (c) (d) (e)

Figure 3.8 Press image of roll press–type nanoimprint. A roll surface has structure (a, b); the foil pattern rotates and is pressed (c, d); and press a foil pattern by roll (e).

From an industrial viewpoint, roll-press nanoimprinting is attractive for mass production. The tact time is drastically reduced in roll pressing compared to parallel pressing because the heat capacity of the mold in roll pressing is smaller than that of the mold in parallel-press imprinting due to the small press area. However, the cooling time is quite short, so controlling the pattern fidelity is difficult.

3.5 Example of Thermal Nanoimprinting: Sheet Nanoimprinting

Practical fabrication of a film by thermal nanoimprinting is explained as follows in terms of examples of processes, materials, molds, and equipment. In this section, continuous film fabrication by roll pressing is explained [3].

3.5.1 Experimental

A schematic diagram of the sheet-nanoimprinting system is shown in Fig. 3.9. The system uses two closed-loop molds, called "belt molds" hereafter. The belt molds are rotated, heated, pressed onto a film, and cooled as they rotate. The thermoplastic film is transferred by a reel-to-reel mechanism. The film is pressed by press rolls, cooled by the belt molds, and rolled up. The basic process of thermal nanoimprinting, namely heating, pressing, cooling, and separation, is continuously executed by just rotating the belt molds in this system.

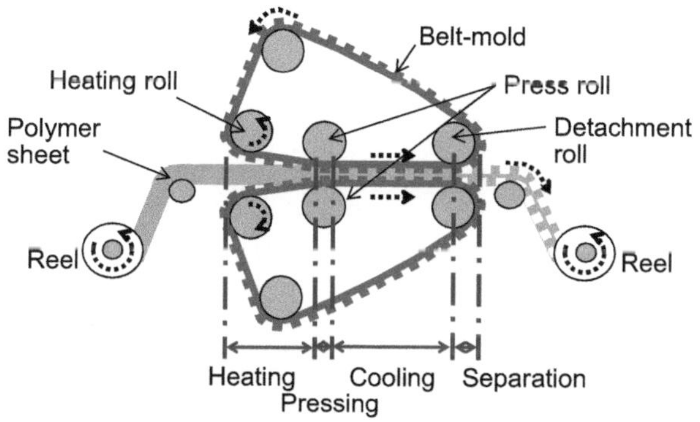

Figure 3.9 Schematic diagram of sheet-nanoimprinting system.

The overall appearance of a sheet-nanoimprinting machine and a close-up shot are shown in Fig. 3.10. The heating rolls and pressing rolls are covered by silicone rubber to compensate for the microroughness of the film and molds.

As shown in Fig. 3.11, the belt mold is composed of a stainless-steel belt and nickel foils, which are formed from a silicon template fabricated by photolithography or EBL. Twenty nickel foils are attached to the stainless-steel belt surface. The nickel foils are obtained by the replication process described in Section 3.3.

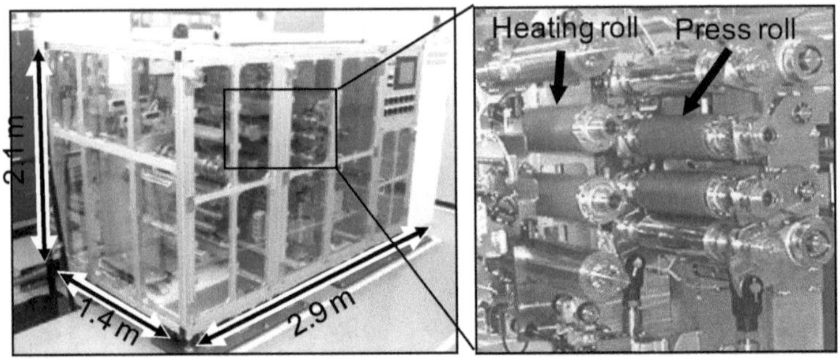

Figure 3.10 Appearance of sheet-nanoimprinting system.

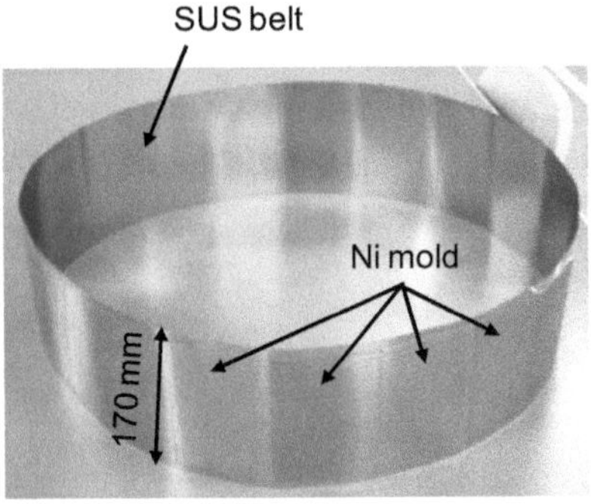

Figure 3.11 Appearance of the belt mold. The Ni molds are mounted on the SUS belt surface.

The nanoimprinting conditions are listed in Table 3.2. The pressure exerted by the rolls is in the typical range of thermal nanoimprinting. The film is polystyrene (PS), with a glass transition temperature of about 100°C. The film speed is 0.3 m/min. Another important process factor is press time. In a sheet-nanoimprinting system, as shown in Fig. 3.12, the press rolls are covered with silicone rubber to generate the appropriate contact width with the PC film. The silicone rubber deforms, as shown in the inserted figure, and a contact width, a so-called nip width, is formed. Under this experimental condition, the nip width was 10 mm. Since the film speed was 0.3 m/min., the press time was 2 s.

Table 3.2 Nanoimprinting conditions

Pressure (MPa)	Temperature (°C)	Film speed (m/min)
1.2	122–148	0.3

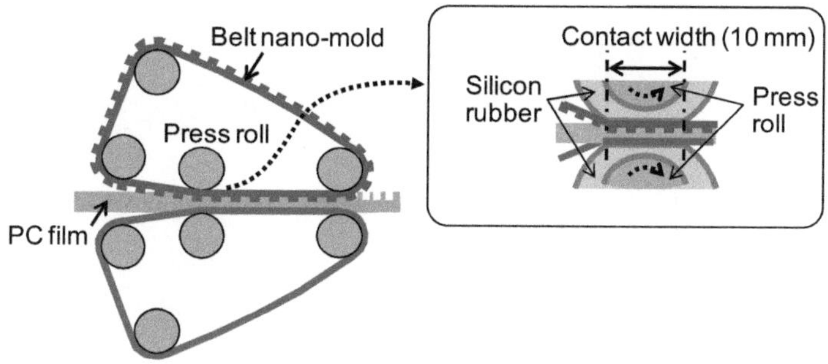

Figure 3.12 Contact status of mold and film in the sheet-nanoimprinting machine. The silicone rubbers of press rolls deform like the inserted figure. The deformed rubbers generate the contact width of pressing for securing the press time.

3.5.2 Results and Analysis

The PS film was imprinted by the sheet-nanoimprinting system under the conditions listed in Table 3.2. An example of an imprinted PS film is shown in Fig. 3.13. The imprinted length is 1.3 m. Several patterns appear on the belt mold, ranging in scale from submicrons to microns, and fine imprinted patterns were confirmed by SEM. It was also demonstrated that the imprinted length reaches the order of kilometers by using the mold-release technique described in Section 2.2.3. The sheet-nanoimprinted 25 nm pillar pattern on the PS film is shown in Fig. 3.14. The imprinted length is 10 m, and 25 nm diameter pillars at three points on the sheet were confirmed by atomic force microscopy (AFM).

The behavior of the resin during thermal nanoimprinting was analyzed. The pillar height is plotted as a function of nanoimprinting temperature (when 400 nm deep holes were used) in Fig. 3.15. Clearly, as the imprint temperature increases, the pillar height increases because the viscosity of the resin is reduced at a high temperature. This resin-filling behavior can be understood by the simple analysis explained as follows.

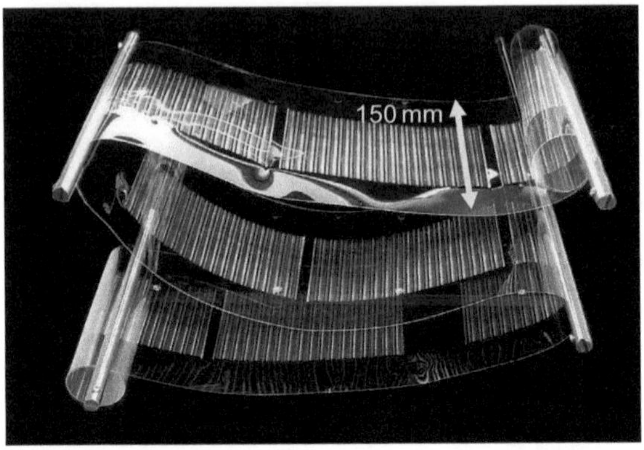

Figure 3.13 Appearance of sheet-nanoimprinted PS film. The imprinted length is 1.3 m, and the film thickness is 80 μm.

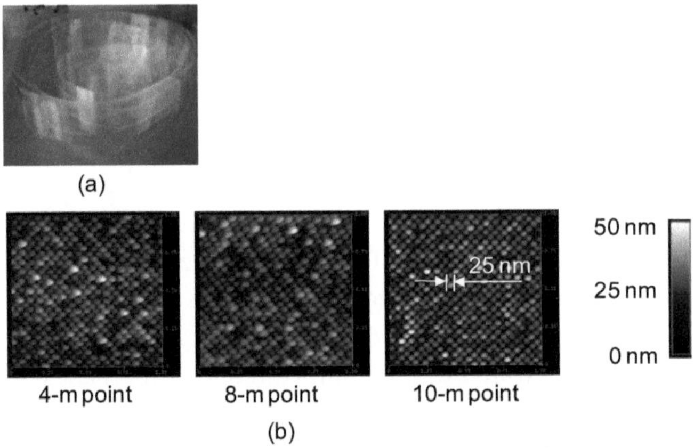

Figure 3.14 (a) Appearance of 10-m-length sheet-nanoimprinted PS film with 25-nm-diameter pillar pattern and (b) AFM observation of three locations of film surface structure. AFM, atomic force microscopy.

A simple model of resin flow into a hole is shown in Fig. 3.16. In general, the filling volume of the viscous flow into a tube is given by the Hagen–Poiseuille equation as

$$V = \frac{\pi r^4 \Delta P}{8LQ}, \tag{3.1}$$

where v is the viscosity of the resin, L is the depth of the mold hole, Q is the flow volume per unit time, r is the radius of the mold hole, and ΔP is the pressure difference between the inlet and the outlet of the tube. The Hagen–Poiseuille equation assumes an open tube, so ΔP indicates the pressure difference for an open tube. The value of ΔP is assumed to be "nanoimprint pressure." Q is given as

$$Q = \frac{\pi r^2 h}{T_\mathrm{p}}, \tag{3.2}$$

where r is the hole radius, h is the pattern height, and T_p is the imprinting time (2 s in this experiment).

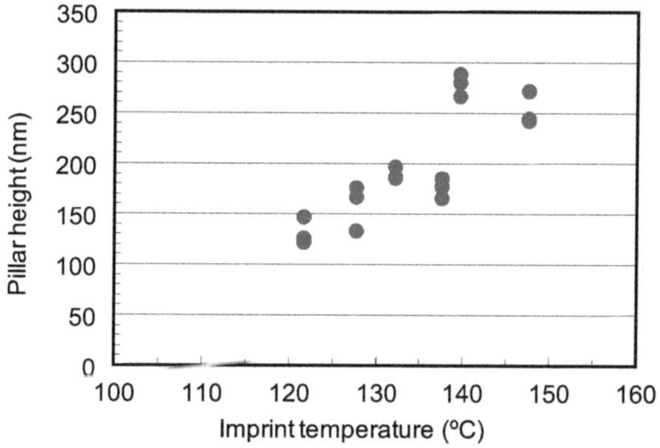

Figure 3.15 Pillar height as a function of imprint temperature. As imprint temperature increases, viscosity of PS decreases, so the pillar height increases.

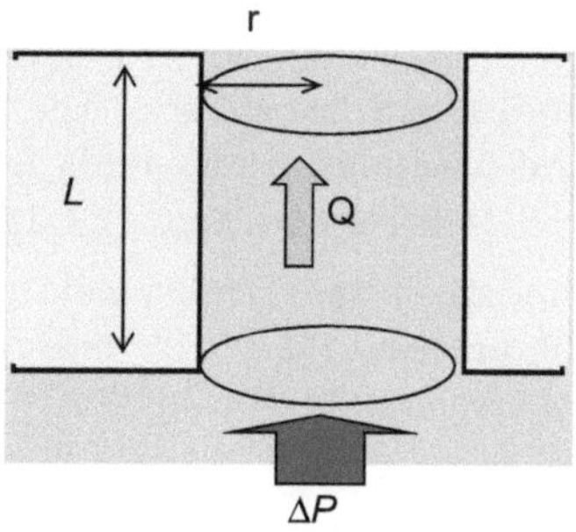

Figure 3.16 Resin flow model. The resin is pressed by pressure difference ΔP and flows into the hole with radius r and depth L. The filled volume of resin per unit time is defined as Q.

The relation between the viscosity of the PS and the imprinted pillar height is shown in Fig. 3.17. Viscosity was controlled by film temperature, that is, the nanoimprinting press temperature.

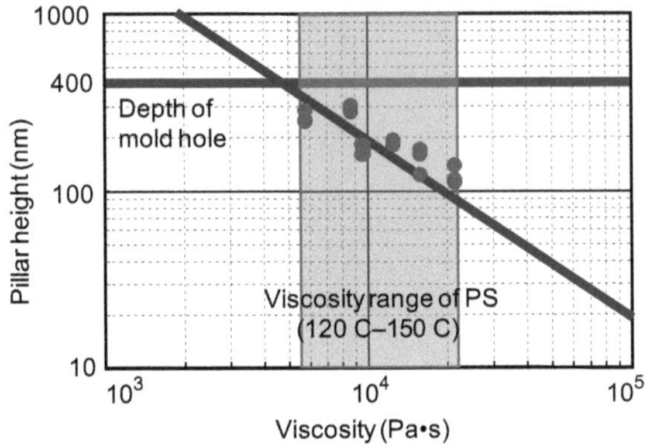

Figure 3.17 Prediction line and experimental plots of pillar heights. The viscosity was controlled by heating roll temperature.

The plots indicate experimentally measured data, and the solid line indicates the predicted pillar height when using the flow model based on the Hagen–Poiseuille equation with r = 100 nm. The measurement plots are on the line given by the Hagen–Poiseuille equation, so it is concluded that the flow behavior of the polymer obeys the Hagen–Poiseuille equation (at least in the submicron-scale space used for the model).

References

1. H. Lim, K.-B. Choi, G. Kim, S. Lee, H. Park, J. Lee, J. Ryu, J. Lee, and S. Jung (2014). Roll-to-roll nanoimprint lithography for patterning on a large-area substrate roll, *Microelectron. Eng.*, **123**, 18–22.
2. N. Ito (2012). Large area seamless roller mold using fast EB lithography (rEBL) for R2R process, *Proc. 11th International Conference on Nanoimprint and Nanoprint Technology*, 7.2.
3. M. Ogino, M. Hasegawa, A. Miyauchi, K. Sakaue, and S. Nagai (2013). Fabrication of 200-nm dot pattern on 15-m-long polymer sheet using sheet nanoimprint method, *Jpn. J. Appl. Phys.*, **52**(3), 035201.1–035201.4.

Chapter 4

Photonanoimprinting

Akihiro Miyauchi
Tokyo Medical and Dental University, 2-3-10 Kanda-Surugadai,
Chiyodaku, Tokyo 101-0062, Japan
miyauchi.ibb@tmd.ac.jp

Photonanoimprinting uses photocurable resins as molding materials. The process temperature is room temperature, so a high fidelity can be expected due to small changes of thermal expansion. In this chapter, photonanoimprinting is explained in terms its process, materials, and equipment. And a practical example of photonanoimprinting is described.

4.1 Process

The basic process of photonanoimprinting is shown in Fig. 4.1. First, the photocurable resin is coated on the substrate. Then, the transparent mold is pressed onto the resin and the resin is cured by ultraviolet (UV) light. After curing, the mold is removed. This sequence is basically the same as thermal nanoimprinting,

Nanoimprint and Its Applications
Edited by Akihiro Miyauchi

ISBN 978-981-4800-37-2 (Hardcover), 978-0-429-03192-2 (eBook)
www.jennystanford.com

except the resin used in photonanoimprinting is UV curable, not a thermoplastic. Since the resin is cured by UV light, the mold or substrate must be transparent in the UV region. In general, glass or polymer materials are used as transparent molds.

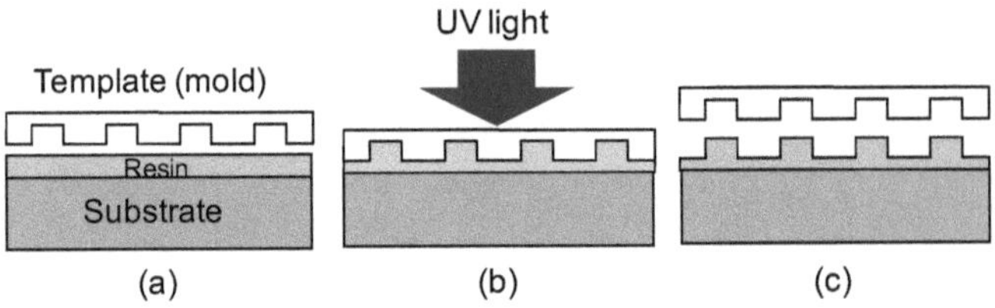

Figure 4.1 Process sequence of photonanoimprinting: (a) the resin layer and template (mold) are prepared, (b) the resin is irradiated with UV light, and (c) the template is released.

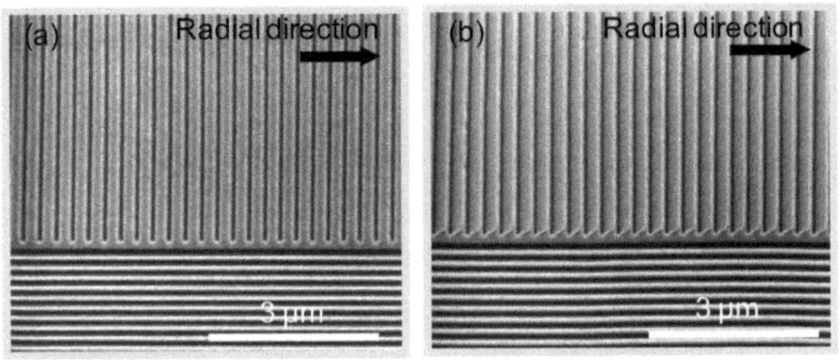

Figure 4.2 Comparison of pattern fidelities produced by (a) photo-and (b) thermal nanoimprinting. The same mold with rectangular grooves was used in both cases. Rectangular lines are formed by photonanoimprinting; in contrast, the lines perpendicular to the radial direction formed by thermal nanoimprinting are deformed as a result of the difference in thermal expansion coefficients of the mold and substrate.

The process temperature of photonanoimprinting is room temperature, so pattern deformation by thermal expansion is quite small. A high fidelity in regard to pattern formation is, therefore, expected. Scanning electron microscopy (SEM) images of photo- and thermal nanoimprints formed on a circular silicon substrate (silicon wafer) are shown in Fig. 4.2. The mold has rectangular grooves, so the imprinted patterns are rectangular lines, as shown in Fig. 4.2a. However, in the case of thermal nanoimprinting, Fig. 4.2b, the lines perpendicular to the radial direction are deformed. This deformation is caused by the difference in the thermal coefficients of the mold and substrate. Photonanoimprinting is thus suitable for fabricating fine devices with high pattern fidelity.

For UV curing, a high-pressure mercury lamp or UV light-emitting diodes (LEDs) are used as the light source. The optical design of a photonanoimprinting system differs from that of photolithography systems. That is, resolution is defined by the mold pattern used in photonanoimprinting, so only light-intensity distribution is considered in the optical design of photonanoimprinting. The cost of nanoimprinting machinery is, therefore, considerably lower than that of photolithography machinery.

4.2 Materials

In photonanoimprinting, the materials used are categorized as UV-curable resins, coupling agents, mold-release agents, and molds. The mold-release materials are described in Section 2.2. The mold materials are explained in Section 4.3. UV-curable resins and coupling agents are described in this section.

4.2.1 UV-Curable Resin

Various kinds of UV-curable resins are commercially available, and the suitable photocurable resin is chosen on the basis of the required device fabrication process.

Characteristics of photocurable monomers are listed in Table 4.1 [1]. Curing is categorized into two main types, radical and cationic. (Note that another type, anion curing, is not mentioned here.) When UV light irradiates the photocurable resin, radicals or acid is generated in the resin by a photoinitiator and a curing reaction occurs. The reaction rate of radical curing is higher than that of cationic curing. However, radicals are easy to deactivate, and cationic reactions continue after UV radiation, so dark reactivity of cationic curing is higher than that of radical curing. Radicals are also deactivated by oxygen, so it is difficult to cure the resin if it is in contact of an oxygen-containing gas such as air.

Table 4.1 Characteristics of UV-curable monomers [1]

	Radical curing	Cationic curing			
	Acryl type	**Glycidyl type epoxy group**	**Alicyclic epoxy group**	**Oxetane group**	**Vinyl ether group**
Reactivity	Good ~Very good	Very poor	Good	Good	Very good
Dark reactivity	Very poor	Good	Good	Good	Good
Heat resistance	Good	Good	Very good	Poor	Very poor
Curing shrinkage	Several tens %	Several %	Several %	Several %	Several tens %
Anti oxygen inhibition	Very poor	Good	Good	Poor [2]	Good
Molecular weight	Very good	Poor	Good	Good [3]	Good
Types	Very various	Poor	Various	Poor	Poor
Viscosity	Low	Very high	High	Low	High
Toxicity	High ~Low	High ~Low	High ~Low	Low	Low

Viscosity is important because the UV-curable resin fills the mold cavities by capillary action. The viscosity of commercially available UV-curable resin for photonanoimprinting is around several mPa·s, so the resin is almost like water in terms of viscosity.

Toxicity is also important. Conventional photocurable resins used in photolithography are removed after etching. Consequently, the cured resins do not remain in the final products, such as large-scale integration (LSI) circuits and liquid-crystal display (LCD) panels. However, in some applications, the cured resins remain and act as devices. For example, in the case of cell cultures in biology, cells are cultivated on nanoimprinted resin structures, so toxicity is a very important factor in choosing nanoimprint resins.

As mentioned before, many kinds of UV-curable resins are commercially available. Each resin has unique characteristics; accordingly, on the basis of the characteristics of monomers in the resin, it is necessary to choose which resin is suitable for fabricating a target device.

4.2.2 Coupling Treatment

The key factor in high-quality nanoimprinting is controlling the release of the mold and resin. The release property is basically governed by the balance of sticking forces between the substrate and the resin and between the resin and the mold. This sticking force balance is explained schematically in Fig. 4.3. When the mold is removed from the resin, the resin has to remain on the substrate, so the sticking force between the resin and the substrate must be higher than that between the mold and the resin. To achieve this sticking force balance, release treatment of the mold and coupling treatment of the substrate are generally used in nanoimprinting. The mold-release treatment is standard in both thermal nanoimprinting and photonanoimprinting.

As for the coupling treatment, the chemical state of the substrate's surface is an important factor due to the reaction between the surface and coupling agents. Substrates for nanoimprinting are generally cleaned by chemical wet treatment before nanoimprinting because the substrate surface is usually contaminated by contaminants such as organics, metals, and tiny particles. In general, substrates are cleaned by using acid, alkali, and organic solvents. The final wet treatment is generally rinsing with ultrapure water. After the ultrapure-water rinse, the substrate surface is covered by hydroxyl groups, with which the coupling agent reacts. An example of a coupling agent, 3-acryloxypropyltrimethoxysilane, is shown in Fig. 4.4. This coupling agent reacts with the surface hydroxyl groups, thereby forming acrylic groups on the substrate surface. The acrylic groups then react with the resist components and form covalent bonds. As a result, as shown in Fig. 4.5, the substrate and the resin molecules are covalently bonded via the coupling agent.

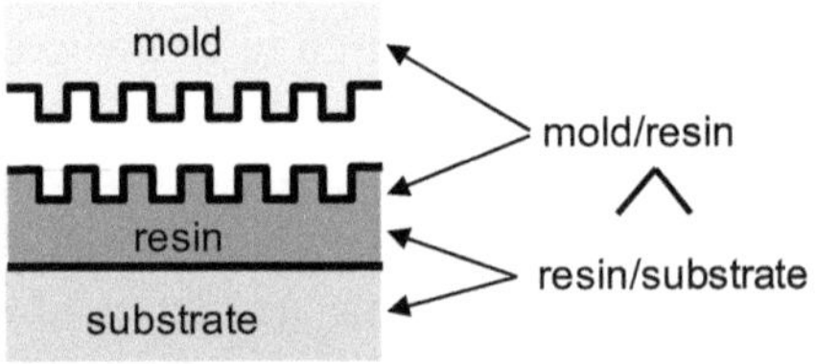

Figure 4.3 Balance of sticking forces between nanoimprint materials. Sticking force between the resin and the substrate must be stronger than that between the mold and the resin.

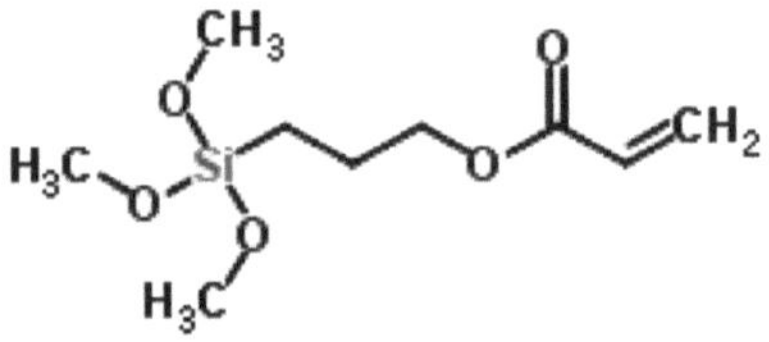

Figure 4.4 Typical coupling molecule: 3-acryloxypropyltrimethoxysilane (ATMS).

The sticking forces between mold and resin and between resin and glass are plotted in Fig. 4.6. The mold-resin sticking force was 0.17 MPa. Accordingly, the resin-glass sticking force must be higher than 0.17 MPa to keep the resin on the glass substrate. The resin-glass sticking force glass was 0.25 MPa, which was not high enough to prevent the resin from peeling off. However, it was considerably increased (to 70 MPa) by the coupling treatment. Accordingly, the resin sticks to the glass substrate with a force 410 times stronger than the force with which it sticks to the mold. The effect of the coupling treatment in a practical case is explained in Section 4.4.

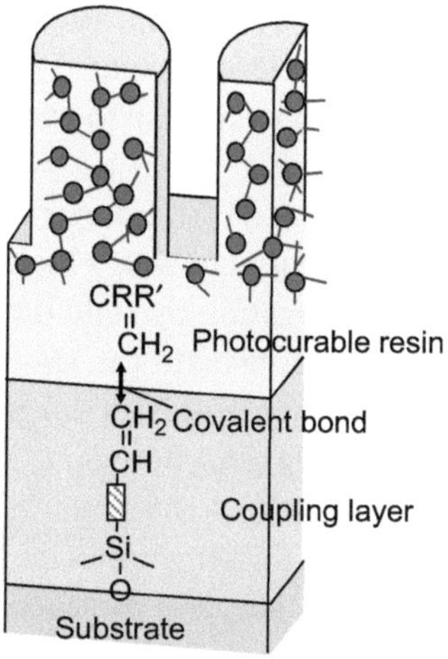

Figure 4.5 Model of chemical bonding of a photocurable resin and a substrate surface via a coupling layer.

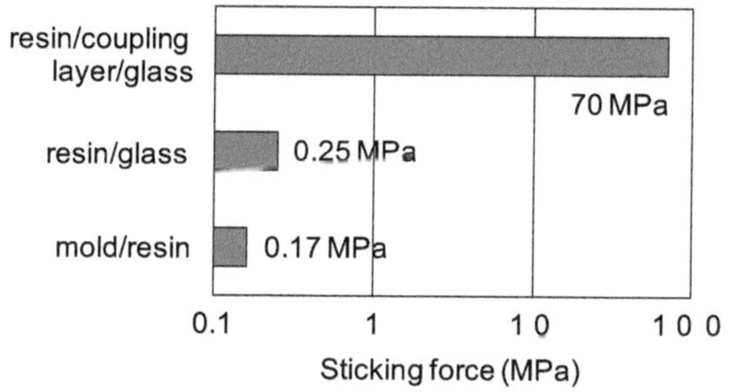

Figure 4.6 Sticking force at each interface. Sticking force between the resin and glass is significantly increased by coupling treatment.

4.3 Molds

In photonanoimprinting, the resin is cured by UV light, so the mold or the substrate must be transparent. Two kinds of mold for photonanoimprinting are described in this section.

4.3.1 Glass Mold

The standard mold for photonanoimprinting is made of glass (often quartz), as shown in Fig. 2.4, for high fidelity. The glass mold is generally fabricated by photolithography or electron beam lithography (EBL). Since glass is an insulator, metals (like chromium) are often deposited on the glass surface to prevent charge-up during EB writing and/or dry etching.

A glass mold (namely quartz) is standard in photonanoimprinting for LSI applications because it achieves a high fidelity. The demerit of a glass mold is defect generation by solid particular contaminants between the mold and the substrate (as shown in Fig. 4.7). When both mold and substrate are solids, a large noncontact area is formed when a solid particle gets between them. This noncontact phenomenon causes nonfilling of the resist into the mold cavities, thereby generating "pattern-missing" defects. Ultraclean technology is, therefore, important for high-yield production using glass molds.

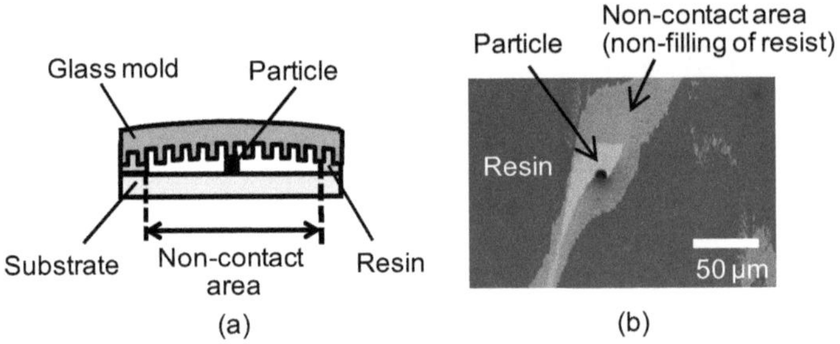

Figure 4.7 Schematic diagram of defect generation by particle contamination. (a) The solid glass mold is hard, so a large noncontact area is formed if a solid particle is inserted between the mold and the solid substrate. (b) Around the particle, the mold cannot contact the substrate, and a nonfilling defect is generated.

4.3.2 Soft Mold

As mentioned before, a noncontact area is generated by a solid contaminant particle when both mold and substrate are rigid. If the mold is soft, like a polymer, the noncontact area can be reduced. A mold made of polymer is called a "soft mold."

A typical structure of a soft mold and the result of nanoimprinting when a solid contaminant particle is present are shown in Figs. 4.8a and 4.8b, respectively. The soft mold consists of a poly(dimethylsiloxane) (PDMS) layer and a glass substrate. As a contamination particle, a 1 μm diameter silica bead was used. The particle causes a noncontact diameter of only 2.2 μm. Reduction of the noncontact area by a soft mold is shown in Fig. 4.9. In this case, pillars are formed in line on the substrate as contaminants. The diameter and height of each pillar is 10 μm and 7.6 μm, respectively. When the imprint pressure is 0.5 MPa (Fig. 4.9a), a noncontact area is formed linearly and the noncontact width is about 104 μm. On increasing the imprint pressure, to 1.0 MPa, the noncontact area is reduced and is limited to the vicinity of each pillar. These results demonstrate that the soft mold effectively reduces the noncontact area formed by solid contaminants between mold and substrate.

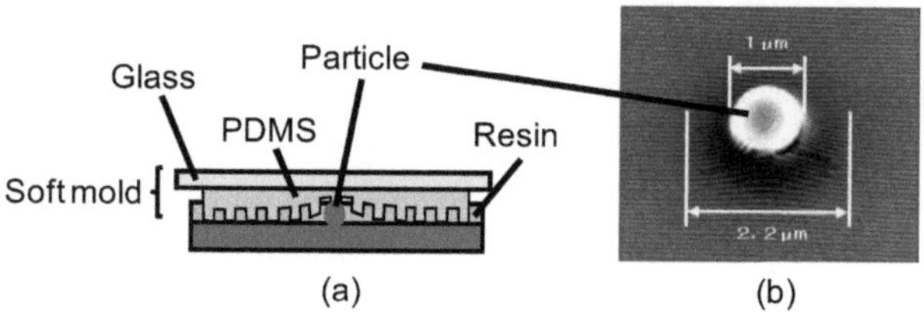

Figure 4.8 (a) Typical structure of a soft mold and a deformation model for the inserted particle. (b) SEM image of the pattern imprinted by using the soft mold. Diameter of the noncontact area is 2.2 μm in the case of a 1-μm-diameter particle. SEM, scanning electron microscopy.

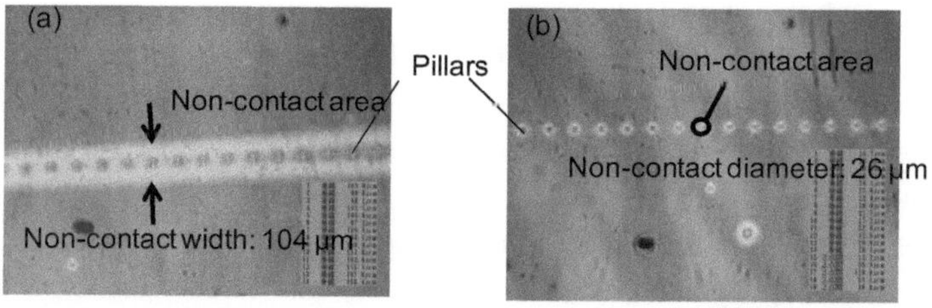

Figure 4.9 Reduction of noncontact area by using a PDMS mold. A line of pillars is formed. The pillar diameter and height are 10 μm and 7.6 μm, respectively. Nanoimprinting pressure is (a) 0.5 MPa and (b) 1.0 MPa. PDMS, poly(dimethylsiloxane).

4.4 Equipment

As with thermal nanoimprinting, two types of photonanoimprinting equipment are mainly used: parallel press and roll press. The difference between the two types in mechanical terms is the pressure-loading mechanics. In thermal nanoimprinting, nanoimprinting pressure is on the order of megapascals. Under such high pressure, air entrapped between the mold and the substrate is compressed and the volume is negligible. However, in photonanoimprinting, the nanoimprinting pressure is low for the glass mold and the resin mainly fills the mold cavity by capillary force, not mechanical pressure. In photonanoimprinting, therefore, the sequence of the pressing process is important to avoid entrapment of air.

4.4.1 Parallel Press

A parallel press is a standard machine for both thermal nanoimprinting and photonanoimprinting. A parallel press for photonanoimprinting mainly consists of press stages and a UV light source. The press stage is driven by oil pressure or servo motor, both of which are suitable for high-pressure and high-precision pressure control, respectively.

A parallel press presses the mold onto a flat substrate at one time, so there is a risk of atmospheric gas getting entrapped. To avoid such gas entrapment, either a chemical or a mechanical approach is taken.

As for the chemical approach, the atmospheric gas around mold and substrate is controlled. It is well known that helium has a small atomic radius and relatively high solubility in monomers. Some photonanoimprinting machines use helium as an atmospheric gas around mold and substrate. Use of liquid gas has been reported to eliminate air bubbles. For example, it was reported that the pentafluoropropane (PFP, $CHF_2CH_2CF_3$) gas effectively eliminates air bubble defects [4]. The vapor pressure of PFP gas is 0.1495 MPa at 25°C; therefore, PFP gas becomes liquid during the pressing process, thereby preventing generation of air bubble defects. Details are described in Appendix C.

As for the mechanical approach, it is important to control contact distribution between mold and substrate. A schematic diagram of the press sequence to avoid air entrapment is shown in Fig. 4.10. The mold is a soft mold, so it is flexible. The soft mold is pressed from its backside by the spherical glass press and bends in a convex shape. Then, the substrate gets attached to the mold from the center to the edge and, finally, the whole mold surface comes in contact with the substrate. Air is pushed out by this action; so as shown in Fig. 4.11, no air is entrapped.

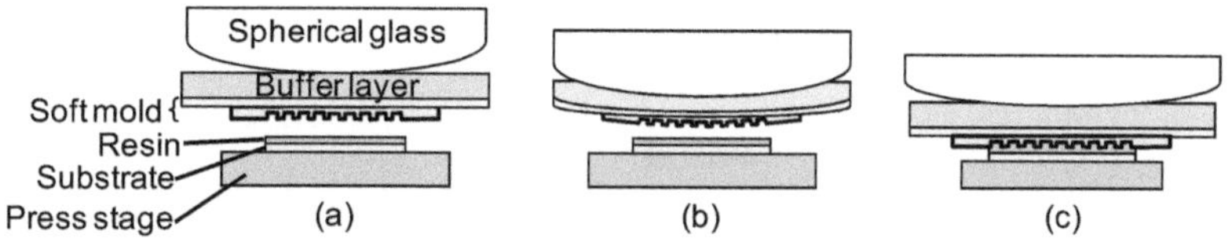

Figure 4.10 Schematic diagram of press sequence for avoiding air entrapping: (a) pressing the buffer layer by a spherical glass press, (b) bending the soft mold to form a convex shape, and (c) contacting the mold and substrate from center to edge.

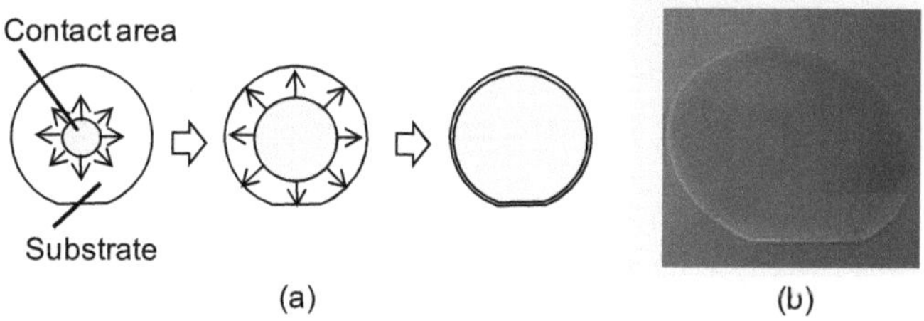

Figure 4.11 (a) Schematic diagram of contact area distribution and (b) appearance of imprinted wafer without air bubbles.

4.4.2 Roll Press

A roll press can also push out atmospheric gas during the pressing. A typical roll system is shown in Fig. 4.12. It is based on a conventional 2P (photopolymerization) roll system. The film is transferred by a reel-to-reel process, as follows. A photocurable resin is coated on the film surface by a die coater. The resin layer is pressed by the roll mold and cured by UV light, so a patterned resin can be obtained continuously. The coated resin layer comes in contact with the roll mold gradually, so no air is entrapped between the resin and the roll mold. A high-pressure mercury lamp or UV LED is used as a light source. Compared with a high-pressure mercury lamp, an LED light source is very small and generates little heat. An LED chip is several millimeters square, so the position of each LED chip is designed by optical simulation to get the required light-intensity distribution.

Compared with the optical design of a photolithography machine, that of a photonanoimprint machine is quite simple, because the pattern fidelity is defined by mold morphology, not by light paths.

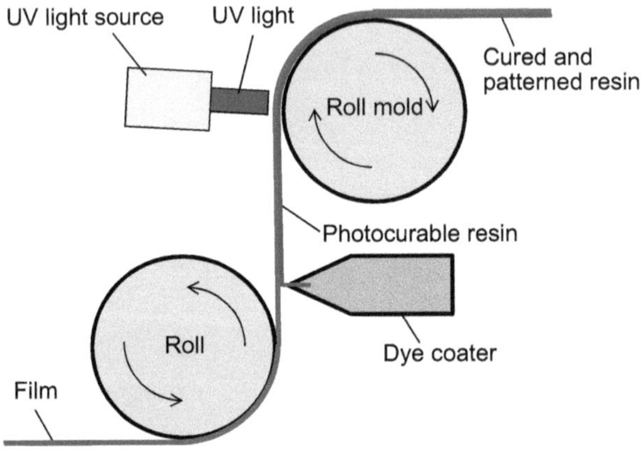

Figure 4.12 Schematic diagram of roll-type photonanoimprinting machine.

4.5 Example of Photonanoimprinting

Photonanoimprinting is suitable for nanoscale patterning. However, several tips concerning successful photonanoimprinting should be considered. In this section, as an example, fabrication of an aluminum film by photonanoimprinting is explained.

4.5.1 Process Sequence

The process sequence of aluminum fabrication as an example of photonanoimprinting is shown in Fig. 4.13. An aluminum layer is deposited on a glass substrate by sputtering or EB evaporation. The thickness of aluminum is 135 nm. The resist is spin coated on the aluminum layer. The resist thickness is 90 nm. The soft mold is pressed onto the resist layer. The soft mold has a groove pattern with a groove depth of 120 nm. After UV curing, the soft mold is demolded. The aluminum layer is then fabricated by dry etching. By this dry etching, the residual layer of the resist is etched away. After the dry etching, the remaining resist is removed.

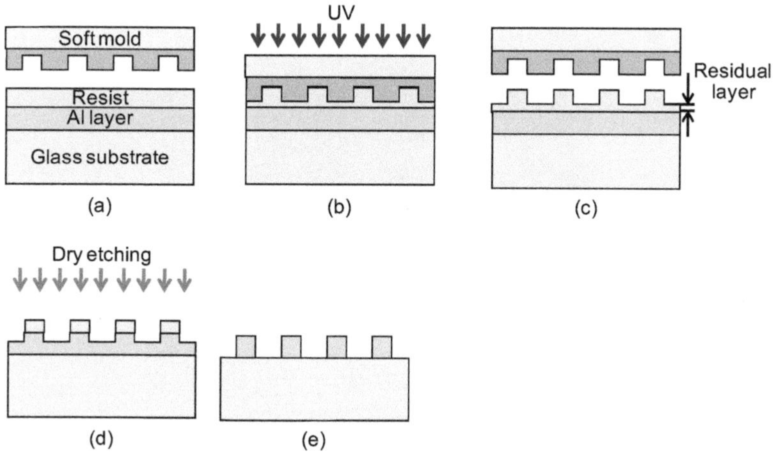

Figure 4.13 Processing sequence for fabricating aluminum by photonanoimprinting: (a) spin-coating the resist on the aluminum/glass substrate, (b) pressing the soft mold and irradiating by UV light, (c) demolding the soft mold after curing, (d) dry-etching the aluminum layer, and (e) removing the resist.

As can be seen from this sequence, the uniformity of the residual layer is important to achieve aluminum wires of a uniform height. The residual layer thickness is determined by the initial volume of the resist and the total cavity volume of the mold groove pattern. If the initial resist volume is not enough to fill the grooves on the mold, faults in the line pattern are generated. On the other hand, if the initial volume of the resist is too much to fill the grooves on the mold surface, the uniformity of the residual layer is lost. Therefore, the spin coating condition is important for forming a uniform residual layer.

4.5.2 Soft Mold

The silicon template for the soft mold is fabricated by EBL, as shown in Fig. 4.14. The diameter of the silicon wafer is 200 mm. The pattern is so-called line and space (L&S). The full pitch, the line width, and the line height are 135 nm, 55 nm, and 120 nm, respectively.

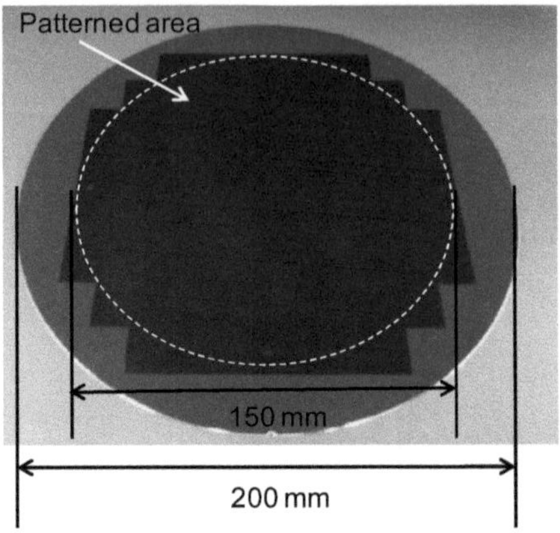

Figure 4.14 Appearance of a silicon template fabricated by EBL. The diameter of the silicon wafer is 200 mm. The dark area is the patterned area (line-and-space pattern). Full pitch, line width, and line height are 135, 55, and 120 nm, respectively.

The substrate of the soft mold is a glass plate, and patterned polymer layers were obtained by photonanoimprinting. The polymer of the soft mold is a cationic curable resin. Here, the resist is usually a radical curing resin for fast curing. It is important to use different curing type of resins to prevent the sticking of soft mold and resist.

4.5.3 Coupling Treatment Process

As mentioned in Section 4.2.2, coupling treatment is important to prevent the patterned resin from peeling. As shown in Fig. 4.6, the conventional coupling treatment is effective on glass or silicon. However, the coupling treatment is insufficient for some kinds of metals or nitrides because the surface density of the hydroxyl group is low. In the case of an aluminum surface, as shown in Fig. 4.15, the coupling effect is insufficient and peeling defects are observed when the agent 3-acryloxypropyltrimethoxysilane (ATMS) is used. Therefore, the coupling treatment is improved as explained below.

To increase the coupling reaction density, 1,2-bis(trimethoxysilyl) ethane (BTE) is inserted between the ATMS and the aluminum surface. BTE has three silane-coupling sites at both ends, as shown in Fig. 4.16. The coupling treatment of an ATMS/BTE/Al substrate restricts the peeling of the resist, as shown in Fig. 4.17. In Fig. 4.17a, the area ratio of peeling defects is plotted as a function of the number of continuous nanoimprint cycles. Clearly, the area of peeling of the resist increases with an increasing number of nanoimprint cycles in the case of ATM/Al coupling treatment. However, after BTE is used, the area of peeling defects is restricted. Fine L&S patterns were observed by SEM observation, as shown in Fig. 4.17b.

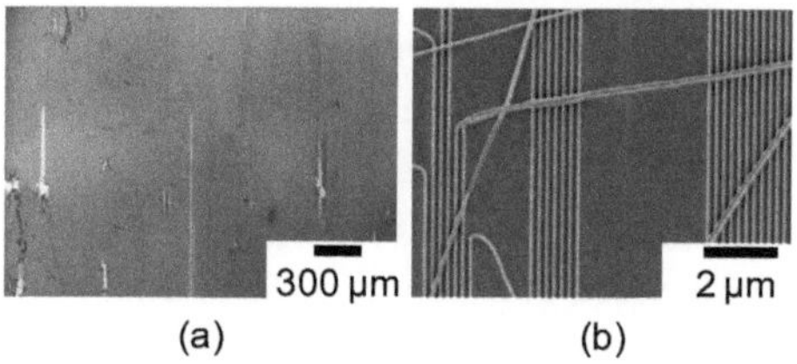

Figure 4.15 Peeling defects observed in conventional coupling treatment: (a) optical microscopy image and (b) SEM image in which the white lines are the peeling regions shown in (a). SEM, scanning electron microscopy.

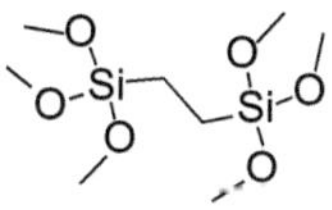

Figure 4.16 Molecular structure of 1,2-bis(trimethoxysilyl) ethane (BTE).

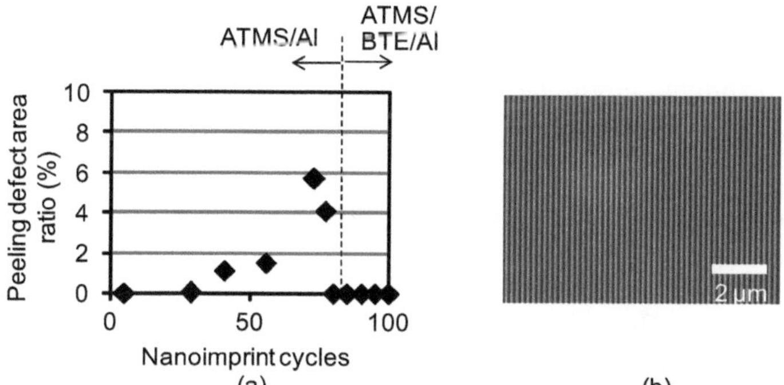

Figure 4.17 Reduction of peeling defects by ATMS/BTE/Al coupling treatment: (a) significant reduction of resist peeling by BTE. A fine line-and-space pattern was observed by SEM. (b) Sample formed by the 90th nanoimprint cycle. ATMS, 3-acryloxypropyltrimethoxysilane; BTE, 1,2-bis(trimethoxysilyl) ethane; SEM, scanning electron microscopy.

As mentioned before, resist peeling is a major problem in photonanoimprinting, so effective coupling treatment is important for reducing the area of peeling defects.

4.5.4 Dry Etching of Aluminum

Fine nanoimprinted resist patterns are formed as shown in Fig. 4.17b. The next step is aluminum dry etching. In this example, the resist residual layer was removed by nitrogen dry etching. After removal of the residual layer, an aluminum layer is etched by using chlorine/BCl_3 mixed gas. Fine dry-etched aluminum nanowires, as shown in Fig. 4.18, were confirmed by SEM observation.

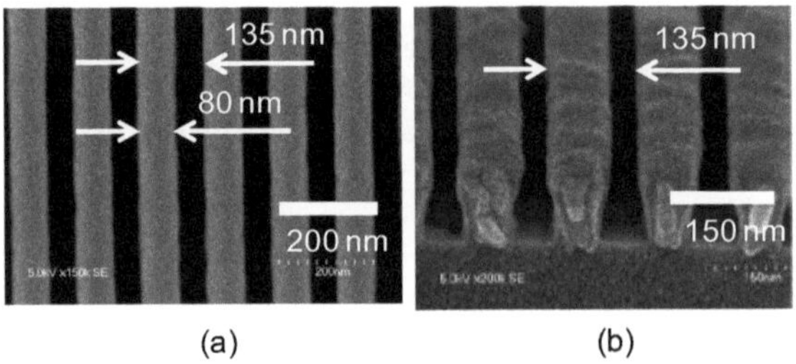

Figure 4.18 SEM images of fabricated aluminum nanowires: (a) plane view and (b) titled cross-sectional view. SEM, scanning electron microscopy.

As described in this section, process problems, such as lack of photocuring, press nonuniformity, and resist peeling, often occur. Accordingly, modification of the process and optimization of materials are important for practical mass production.

Appendix A: Fidelity of a Soft Mold

As described in Section 4.3.2, a soft mold effectively reduces the defect area. One of the concerns is the fidelity of pattern formation by a soft mold because the hardness of the mold pattern is less than that formed by a mold made of glass, nickel, or silicon. The fidelity of a soft mold was investigated as described below. It was evaluated as the line edge roughness (LER) of line patterns.

A master silicon mold was fabricated by anti-isotropic etching of crystalline silicon. The etching rate of the crystal silicon depends on crystal orientation, so a (111) vertical sidewall is formed by using a

(110) silicon wafer. The L&S pattern was fabricated by anti-isotropic etching using a (110) silicon wafer, as illustrated in Fig. 4.19. The pitch, the line width, and the height of the prepared silicon template are 100 nm, 50 nm, and 105 nm, respectively.

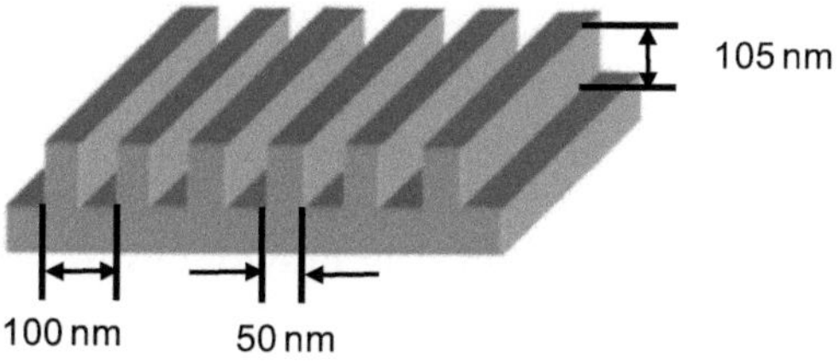

Figure 4.19 Schematic diagram of a silicon template prepared by anti-isotropic etching.

Plane SEM images of a master template (anti-isotropic etched silicon), a soft mold, and an imprinted resist are shown in Table 4.2. The features of the master template and the imprinted resist are identical because of the twice-nanoimprint process. The LER and the line width roughness (LWR) were evaluated by critical dimension (CD) SEM. The LER 3σ of the master template and the imprinted resist were 1.5 and 1.9 nm, respectively. The degradation of LER 3σ is thus 0.4 nm. In other words, degradation of the LER when a soft mold is used becomes atomic level.

Table 4.2 Degradation of line edge roughness when a soft mold is used

	Master template	Soft mold	Imprinted resist
SEEEM images	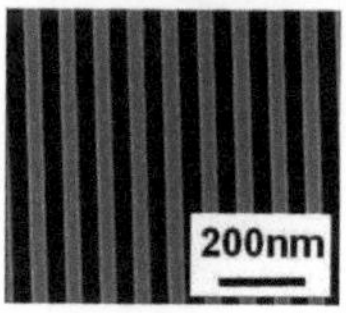		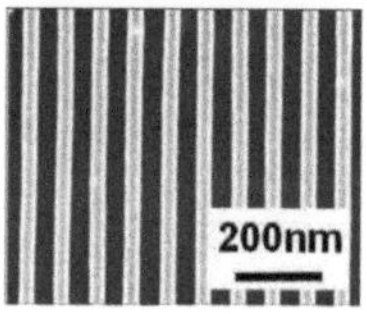
LW	47.9 m	—	47.0 nm
LWR3σ	1.5 nm	—	2.0 nm
LWR3σ	1.5 nm	—	1.9 m

The pitch fidelity has been evaluated by high-resolution grazing-incidence X-ray diffraction (HRGIXD) [5]. The samples were the same as those shown in Table 4.2. X-ray-scattering spectra from a silicon template and an imprinted resist are shown in Fig. 4.20. The oscillations in the spectra were generated by interference of reflected X-rays from sidewall structures. Therefore, the pitch of the walls can be calculated from the oscillation periods.

The q value (plotted in Fig. 4.21) is defined as

$$q = \frac{4\pi \sin\theta}{\lambda}, \tag{4.1}$$

where θ is the incident angle of the X-ray to the sample and λ is the wavelength of the X-ray beam. The fidelity of the line pitch can be calculated on the angstrom scale by HRGIXD, and the pitch of the imprinted resist pattern is identical to that of the master template.

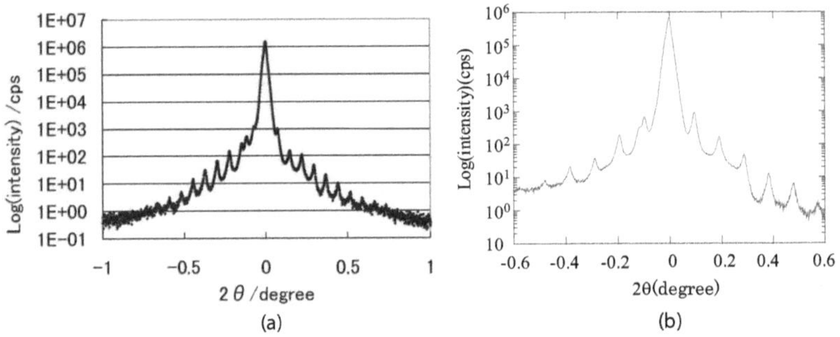

Figure 4.20 X-ray scattering spectra: (a) silicon template and (b) imprinted resist.

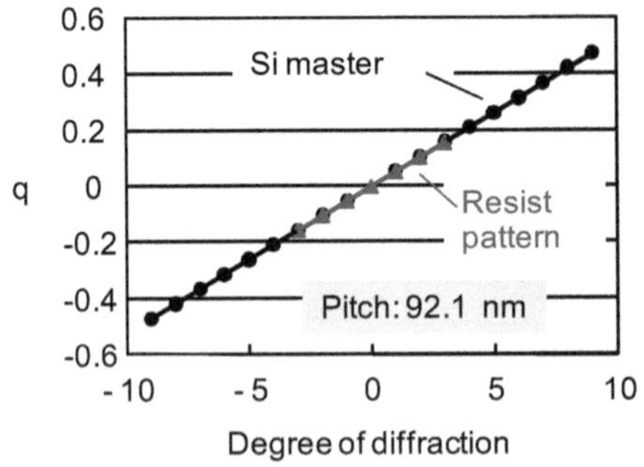

Figure 4.21 Degree of diffraction.

As described in this appendix, fidelities of nanoimprinted patterns formed using a soft mold were evaluated by CD SEM and HRGIXD. These measurements indicate that pattern degradation due to the soft mold is less than a few angstroms. It is, therefore, concluded that a reasonably high fidelity can be obtained by using a soft mold.

Appendix B: Resin Viscosity in Nanospace

As described in Section 4.5, uniformity of the residual layer thickness is a key factor in dry etching. A residual layer is formed as a result of resin flowing between mold and substrate. In photonanoimprinting, as shown in Fig. 4.22, the residual thickness is several to several dozen nanometers. Therefore, sufficient resin flow in the nanospace is crucial for forming a uniform residual layer. In this appendix, the behavior of a resin in the nanospace is discussed in view of viscosity.

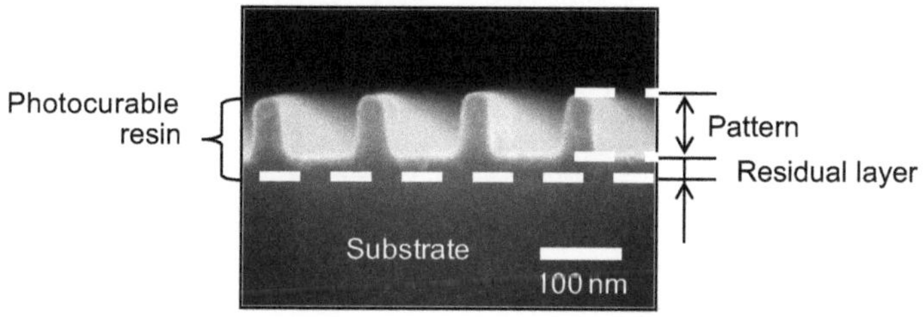

Figure 4.22 Typical image of a residual layer in photonanoimprinting. A residual layer is formed as a result of resin flow between the mold and the substrate.

The viscosity of a resin in the nanospace has been investigated by the resonance shear measurement [6]. A schematic diagram of a resonance-shear apparatus is shown in Fig. 4.23. (The details of the equipment and the analysis method are reported in Ref. [7].) The resin to be measured is inserted between two mica sheets. Each mica sheet is glued on a semicylindrical silica lens. The thickness of the resin can be measured from fringes of equal chromatic order (FECO). FECO are generated by the interference between the resin layer, mica sheets, and the semicylindrical lens. The apparatus has upper and lower stages. The lower stage can move upward and make the resin thin. The force applied to the resin is measured by monitoring the bending of the cantilever. A piezo tube (not shown in the figure) is equipped at the upper stage, and the upper stage vibrates. When the frequency of the piezo tube is swept, the upper stage resonates at a certain frequency. The resonance frequency can be obtained by detecting the oscillation amplitude of the upper stage. A schematic diagram of the resonance is shown in Fig. 4.24. Peak (a) is the resonance peak of the upper stage; peak (b) indicates the resonance peak when the upper stage comes in contact with the lower stage. These two peaks are separated because the resonance conditions (such as mass and spring constants) differ when the upper stage is isolated and when it comes in contact with the lower stage.

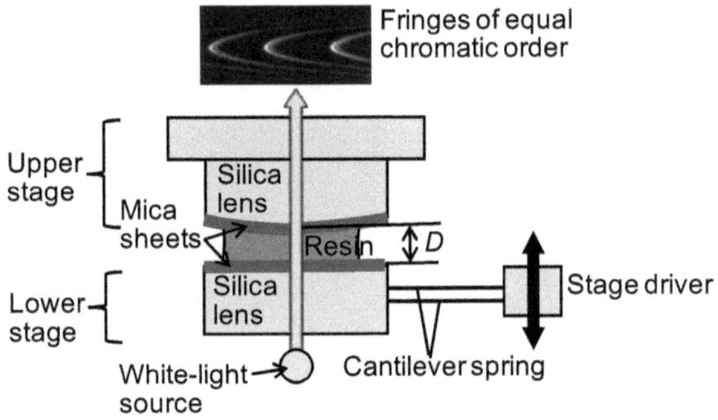

Figure 4.23 Schematic diagram of resonance shear apparatus. The resin sample is inserted between two mica sheets. The thickness of the resin is measured by FECO. As the lower stage moves upward, the resin gets thinner. FECO, fringes of equal chromatic order.

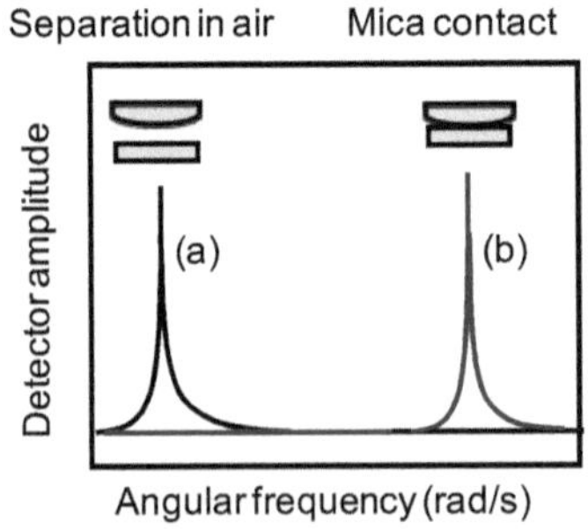

Figure 4.24 Schematic diagram of resonance shear equipment. The resin sample is inserted between two mica sheets. The thickness of the resin can be measured by FECO. The lower stage moves upward, and the resin becomes thin. FECO, fringes of equal chromatic order.

The resin used for photonanoimprinting is inserted between the upper and the lower stages. The viscosity of the resin is about 4 mPa·s (the viscosity of water is about 1 mPa·s), so the vibration of the upper stage does not transmit to the lower stage. As the resin layer gets thinner, the upper and lower stages come in contact and the resonance peak shifts from (a) to (b). Therefore, the viscosity of a resin in nanospace can be measured by monitoring the shift of the oscillation peak.

Measured resonance peaks of the resin are plotted in Fig. 4.25. The oscillation amplitude at 200 rad/s started to diminish when the resin thickness became less than 280 nm. The resonance peaks at each thickness of resin were shifted to 600 rad/s, and when the resin thickness became less than 100 nm, the peaks approached 600 rad/s.

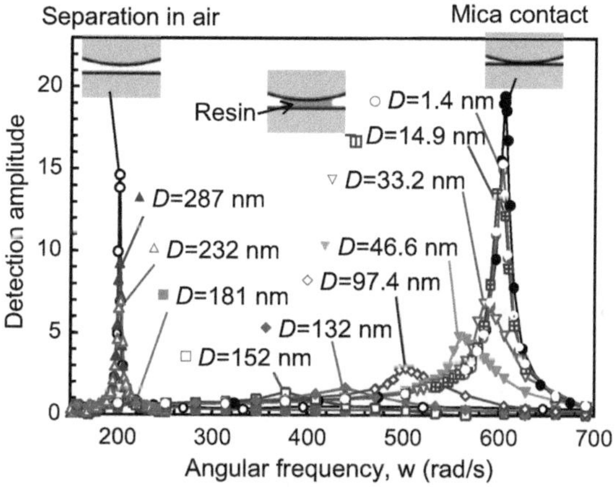

Figure 4.25 Shift of resonance peak during resin thinning. The peak shift occurs at around 200 nm of the distance *D* (thickness of resin), and the peak shifts to contacting mode under *D* less than 100 nm.

The calculated resin viscosity in the nanospace is plotted in Fig. 4.26.

Clearly, the resin viscosity increases when the distance *D*, namely the resin thickness, becomes less than about 150 nm, and it eventually reaches about 20 Pa·s. The bulk viscosity of this resin is about 4 mP·s, so the resin viscosity in the nanospace becomes 5000 times higher than the bulk viscosity. As for the increase of viscosity, it has been reported that water viscosity increases when the water layer thickness becomes less than 3 or 8 nm (depending on the chemical state of the mica surface) due to the structuring of water molecules [8]. The resin viscosity increases in the nanospace seemingly because of the loss of freedom of monomer molecules due to spatial confinement.

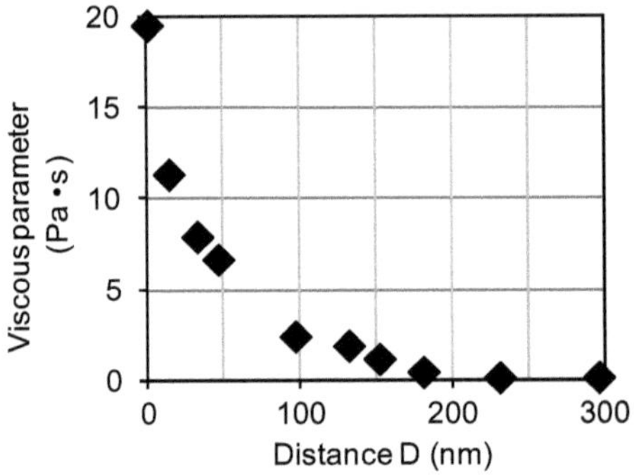

Figure 4.26 Calculated viscosity as a function of distance *D*. Viscosity increases as *D* becomes less than about 100 nm.

Appendix C: Restriction of Air Bubbles[a]

When nanoimprinting is carried out in air, recesses of a mold naturally tend to trap air. The trapped air certainly degrades pattern quality for UV nanoimprinting as the air inevitably remaining as bubble defects although the air hardly impacts patterns fabricated by thermal nanoimprinting. The bubble defects are inherent in UV nanoimprinting. Therefore, special techniques against bubble defects must be used in UV nanoimprinting.

It had been suggested that thermal nanoimprinting should be done in a vacuum [9]. However, it was later claimed that the application of a vacuum was not found to be necessary [10]. The trapped air in the process of thermal nanoimprinting is compressed to a fraction of its original volume, resulting in trapped gas pressure much higher than the original pressure (i.e., atmosphere). Thus, the air can be easily dissolved into the molding material, thanks to

[a]Appendix C has been provided by Hiroshi Hiroshima Research Center for Ubiquitous MEMS and Micro Engineering, National Institute of Advance Industrial Science and Technology (AIST), Tsukuba East, 1-2-1 Namiki, Tsukuba, Ibaraki, 305-8568, Japan Hiroshima-hiroshi@aist.go.jp

Henry's law [11]. It is known that even for thermal nanoimprinting, when the imprint pressure is released before cooling, the air displays an explosive behavior and a viscous fingering pattern, which prefers to grow along the structure edges in a mold, is formed [12].

The imprinting pressure of UV nanoimprinting is 1 to 2 orders of magnitude smaller than that of thermal nanoimprinting and is comparable with ambient atmospheric pressure. Therefore, UV nanoimprinting cannot get the benefits of Henry's law and the bubble defect problem is inherent for UV nanoimprinting in air. The bubble defect problem was hardly mitigated by increasing the imprint pressure several times; however, interestingly, it was drastically mitigated by increasing UV resin thickness [13]. In the latter case, the expected mechanism of bubble defect mitigation is presumably by transportation of bubbles by the massive flow of resin effusing out of a mold. In negligible resin effusing conditions (i.e., when the resin is thin), bubble defects are unavoidable for UV nanoimprinting conducted in air.

The most natural solution against the bubble defect is to reduce environmental air pressure (i.e., using a vacuum) [14]. It was demonstrated that UV nanoimprinting in 1/50 atmosphere always shows excellent results. Using a vacuum is a sure method. However, it detracts from the strong point (i.e., nanofabrication at a low cost) of UV nanoimprinting, as vacuum generation systems are generally expensive. It is necessary to conduct the nanoimprinting process in a nonvacuum environment for preserving its strong point.

UV nanoimprinting named J-FIL [15] (the original name was S-FIL [16]) using a He gas environment [17, 18] combined with dispensing UV-curable resin droplets by ink-jetting is a viable solution for UV nanoimprinting in a nonvacuum environment. In the imprinting process, resin droplets merge with each other [19], preventing He from being trapped during the molding process, although small amounts of He trapped in bubbles left behind are dissolved in the UV-curable resin and diffused into a quartz mold. This method works well, but it might not be compatible with uniformly precoated UV-curable resins prepared by spin coating [20].

A unique UV nanoimprinting process using a condensable gas environment was proposed, and the effectiveness was demonstrated. The first UV nanoimprinting demonstration was carried out using trichlorofluoromethane (CFC-11) [21], and afterward

1,1,1,3,3-pentafluoropropane (HFC-245fa; PFP) [4] was used to address the ozone destruction problem. Table 4.3 summarizes the properties of CFC-11 and HFC-245fa (PFP). The most important characteristic of those gases for the UV nanoimprinting process is the saturated vapor pressure at room temperature. The values are slightly higher than atmospheric pressure, namely, these gases are easily condensed (liquefied) by a little bit of additional pressure during imprinting.

Table 4.3 Properties of 1,1,1,3,3-pentafluoropropane and trichlorofluoromethane

	1,1,1,3,3-Pentafluoropropane (HFC-245fa)	**Trichlorofluoromethane (CFC-11)**
Molecular formula	$CHF_2CH_2CF_3$	CCl_3F
Molecular weight	134	137.37
Boiling point (°C)	15.3	23.85
Vapor pressure at 25°C (MPa)	0.1495	0.1056
Flammability limits in air	None	None
Flammability limits in air	None	None
Toxicity	Low	Low
Global warming potential	1030	4750
Ozone depletion potential	0	1

Figure 4.27 shows the principle of bubble elimination in UV nanoimprinting using a condensable gas. At the beginning of contact between the mold and the UV-curable resin, each recess of a mold traps the condensable gas. The trapped gas is then compressed by the mold. At first, the gas pressure increases inversely proportional to the gas volume. However, the relationship breaks down and condensation (liquefaction) starts when the gas pressure reaches its saturated vapor pressure (e.g., 0.15 MPa for PFP). Finally, the entire trapped gas is completely liquefied. In the case of PFP, the liquid volume turns out to be 1/200 of the gas volume.

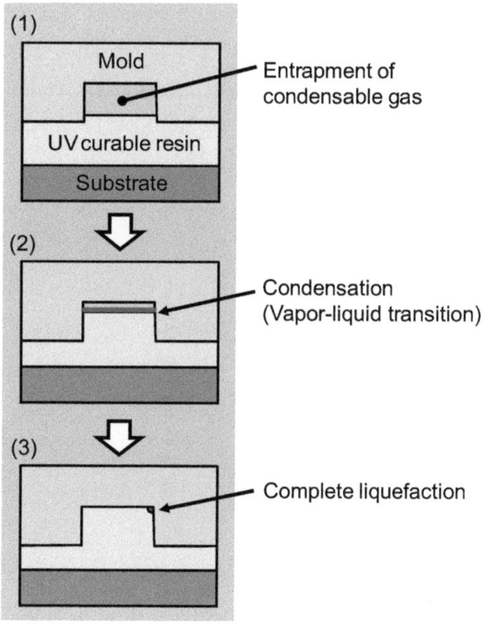

Figure 4.27 Principle of bubble elimination in UV nanoimprinting using a condensable gas.

Figure 4.28 shows cured resin patterns with widths of 2, 3, 5, and 10 μm obtained by UV nanoimprinting in PFP ambience, where a 900-nm-thick spin-coated UV-curable resin (PAK-01, Toyo Gosei) on a Si wafer was imprinted at an imprint pressure of 0.5 MPa by a mold (NIM-PH-350, NTT-AT) containing patterns with widths of 350 nm to 10 μm and a depth of 350 nm. In the experiment, PFP was introduced into the imprinting space at a flow rate of 3000 sccm and the estimated PFP concentration in the imprinting space is approximately 90%. No anomalies are observable in Fig. 4.28. A UV nanoimprinting pattern fabricated in the same imprinting conditions except for PFP introduction (i.e., imprint in air) is shown in Fig. 4.29. Large bubbles exist in 2–3 μm pattern areas, and small bubbles exist in almost every single 10 μm pattern. By comparing Figs. 4.28 and 4.29, it is clearly seen that using PFP as the environment of the UV nanoimprinting process is quite effective. One might suspect that small liquid PFP droplets would create bubble defects as illustrated in Fig. 4.27(3), but in reality, no such defects have been observed in Fig. 4.28. We think that PFP dissolves in UV-curable resin, or it remains as a thin layer between the mold and the UV-curable resin. It is also confirmed that 45-nm-wide L&S patterns can be fabricated by UV nanoimprinting using PFP [22].

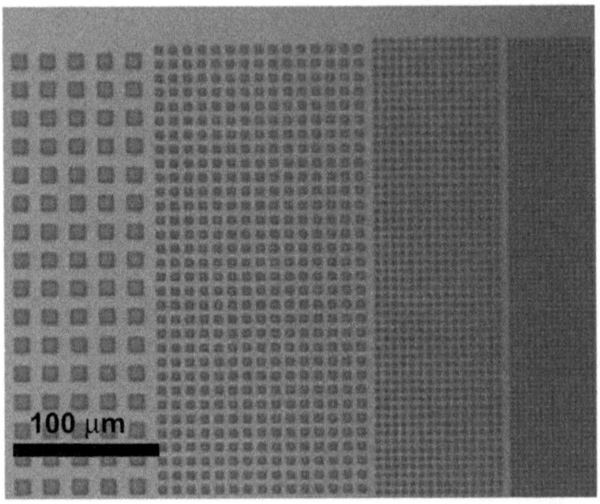

Figure 4.28 Cured resin patterns with widths of 2, 3, 5, and 10 μm obtained by UV nanoimprinting in PFP ambience, where a 900-nm-thick spin-coated UV-curable resin on a Si wafer was imprinted at an imprint pressure of 0.5 MPa by a mold containing patterns with widths of 350 nm–10 μm and a depth of 350 nm. PFP, pentafluoropropane.

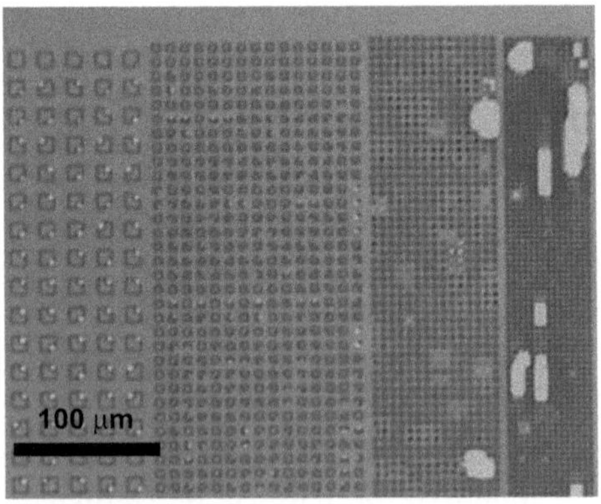

Figure 4.29 UV nanoimprinting pattern fabricated in the same imprinting conditions as Fig. 4.28, except for PFP introduction. PFP, pentafluoropropane.

References

1. H. Miyake (2015). *Nanoimprint Technology*, eds. S. Matsui, and Y. Hirai, Chapter 4: UV nanoimprint (The Institute of Electronics, Information and Communication Engineers, Tokyo), pp. 95–103 (in Japanese).
2. H. Sasaki, A. Kuriyama, and T. Kakuchi (1995). Photoinitiated cationic polymerization of 3-ethyl-3-(phenoxymethyl)-oxetane in nitrogen atmosphere, *J. Macromol. Sci. Pure Appl. Chem.*, **32**(10), 1699–1707.
3. H. Sasaki, J. M. Rudzinnski, and T. Kakuchi (1995). Photoinitiated cationic polymerization of oxetane formulated with oxirane, *J. Polym. Sci., Part A: Polym. Chem.*, **33**(11), 1807–1816.
4. H. Hiroshima, and M. Komuro (2007). Control of bubble defects in UV nanoimprint, *Jpn. J. Appl. Phys.*, **46**, 6391.
5. K. Omote, et al. (2009). A new x-ray metrology for profiling nanostructures of patterned media, *Proc. SPIE*, **7488**, 74881T-1.
6. Y. Shimazaki, S. Oinaka, S. Moriko, K. Kawasaki, S. Ishii, M. Ogino, T. Kubota, and A. Miyauchi (2013). Reduction in viscosity of quasi-2D-confined nanoimprint resin through the addition of fluorine-containing monomers: shear resonance study, *Appl. Mater. Interface*, **5**(16), 7661.
7. M. Mizukami, and K. Kurihara (2008). A new physical model for resonance shear measurement of confined liquids between solid surfaces, *Rev. Sci. Instrum.*, **79**, 113705.
8. M. Kasuya, M. Hino, H. Yamada, M. Mizukami, H. Mori, S. Kajita, T. Ohmori, A. Suzuki, and K. Kurihara (2013). Characterization of water confined between silica surfaces using the resonance shear measurement, *J. Phys. Chem.*, **117**, 13540–13546.
9. S. Y. Chou, P. R. Krauss, and P. J. Renstrom (1996) Nanoimprint lithography, *J. Vac. Sci. Technol. B Microelectron. Nanom. Struct.*, **14**, 4129.
10. L. Heyderman, H. Schift, C. David, J. Gobrecht, and T. Schweizer (2000). Flow behaviour of thin polymer films used for hot embossing lithography, *Microelectron. Eng.*, **54**, 229–245.
11. W. Henry (1803). Experiments on the quantity of gases absorbed by water, at different temperatures, and under different pressures, *Philos. Trans. R. Soc. London*, **93**, 29–274.
12. H. Schift, L. J. Heyderman, J. Gobrecht, and M. A. der Maur (2001). Pattern formation in hot embossing of thin polymer films, *Nanotechnology*, **12**, 173.

13. H. Hiroshima, M. Komuro, N. Kasahara, Y. Kurashima, and J. Taniguchi (2003). Elimination of pattern defects of nanoimprint under atmospheric conditions, *Jpn. J. Appl. Phys., Part 1 Regul. Pap. Short Notes Rev. Pap.*, **42**, 3849–3853.

14. A. Fuchs, M. Bender, U. Plachetka, U. Hermanns, and H. Kurz (2005). Ultraviolet-based nanoimprint at reduced environmental pressure, *J. Vac. Sci. Technol. B Microelectron. Nanom. Struct.*, **23**, 2925.

15. G. M. Schmid, C. Brooks, Z. Ye, S. Johnson, D. LaBrake, S. V. Sreenivasan, and D. J. Resnick (2009). Jet and flash imprint lithography for the fabrication of patterned media drives. In *SPIE Photomask Technology*, p. 748820, International Society for Optics and Photonics.

16. M. Colburn, S. C. Johnson, M. D. Stewart, S. Damle, T. C. Bailey, B. Choi, M. Wedlake, T. B. Michaelson, S. V. Sreenivasan, and J. G. Ekerdt (1999). Step and flash imprint lithography: a new approach to high-resolution patterning. In *Emerging Lithographic Technologies III*, pp. 379–390, International Society for Optics and Photonics.

17. K. Kim, J. Jeong, Y. Sim, and E. Lee (2006). Minimization of residual layer thickness by using the optimized dispensing method in S-FIL TM process, *Microelectron. Eng.*, **83**, 847–850.

18. K. Usuki, S. Wakamatsu, T. Oomatsu, and K. Kodama (2011). Approaches to rapid resist spreading on dispensing based UV-NIL. In *SPIE Advanced Lithography*, pp. 79700S–79700S, International Society for Optics and Photonics.

19. S. Reddy and R. T. Bonnecaze (2005). Simulation of fluid flow in the step and flash imprint lithography process. *Microelectron. Eng.*, **82**, 60–70.

20. R. Kirchner, L. Nüske, A. Finn, B. Lu, and W.-J. Fischer (2012). Stamp-and-repeat UV-imprinting of spin-coated films: pre-exposure and imprint defects, *Microelectron. Eng.*, **97**, 117–121.

21. H. Hiroshima, M. Komuro, N. Kasahara, Y. Kurashima, and J. Taniguchi (2003). Elimination of pattern defects of nanoimprint under atmospheric conditions, *Jpn. J. Appl. Phys.*, **42**, 3849–3853.

22. H. Hiroshima, Q. Wang, and S. W. Youn (2010). 45 nm hp line/space patterning into a thin spin coat film by UV nanoimprint based on condensation, *J. Vac. Sci. Technol. B*, **28**, C6M12-C6M16.

Chapter 5

Room-Temperature Nanoimprinting

Shinji Matsui
Laboratory of Advanced Science and Technology for Industry,
University of Hyogo, 3-1-2 Koto, Kamigori, Ako, Hyogo 678-1201, Japan
shinjimatsui1@gmail.com

We have developed room-temperature nanoimprinting lithography (RT-NIL), which is direct nanoimprinting using sol-gel materials. RT-NIL does not require a resist thermal cycle and UV exposure when pressing a mold onto the resist. Therefore, RT-NIL offers us increased throughput and pattern accuracy in nanostructures fabrication. RT-NIL by a poly(dimethyl siloxane) (PDMS) mold using hydrogen silsesquioxane (HSQ) as a replicated sol-gel material is reviewed, in which the patterning process, 3D structure fabrication, and applications of moth eye structure and gold particle formation for SERS are described.

5.1 Introduction

Nanoimprint lithography (NIL) was proposed by the Chou group in 1995 [1]. NIL has emerged as one of the most promising technologies

Nanoimprint and Its Applications
Edited by Akihiro Miyauchi

ISBN 978-981-4800-37-2 (Hardcover), 978-0-429-03192-2 (eBook)
www.jennystanford.com

for high-throughput nanoscale patterning. The high-resolution and high-throughput capability of NIL makes it a very useful technique in many device applications that require precision patterning of large areas of nanoscale structures. Figure 5.1 shows three types of NIL: thermal NIL (T-NIL) [1], ultraviolet NIL (UV-NIL) [2], and room-temperature NIL (RT-NIL) [3].

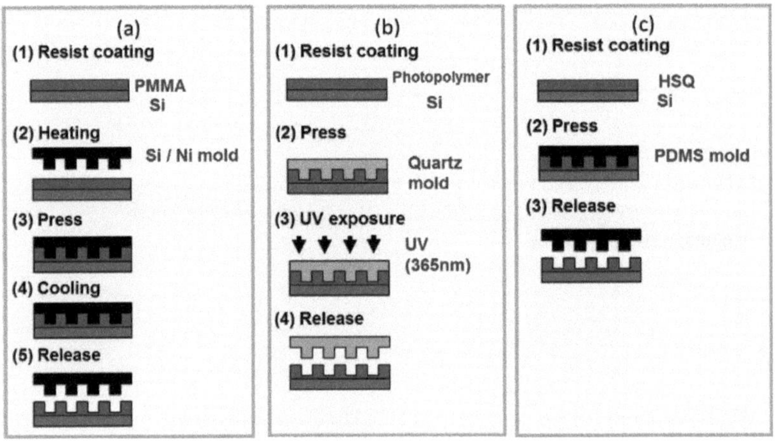

Figure 5.1 Nanoimprinting process: (a) thermal nanoimprint lithography (T-NIL), (b) ultraviolet nanoimprint lithography (UV-NIL), and (c) room-temperature nanoimprint lithography (RT-NIL).

Thermoset or thermoplastic resins are used in T-NIL, requiring a heating and cooling cycle to either cure or melt the resins. When a thermoset resin is used, the mold is pressed on a substrate coated with the resin at room temperature. During pressing, the mold and substrate are heated to harden or cure the thermoset resin, and after cooling, the mold is separated from the substrate. It is slightly different with thermoplastic resin: in this case, the mold is pressed on a substrate coated with the resin held at slightly over the resin glass transition temperature (T_g). The mold and substrate temperatures are then decreased, and the mold is removed from the substrate. Si, SiO_2/Si, and Ni molds are usually used as T-NIL molds. UV-NIL is a room-temperature process because UV-curable resins are used. The UV-NIL mold is pressed onto a substrate coated with UV-curable resin and then the substrate is irradiated with 365 nm UV through the mold. After irradiation, the mold is separated from the substrate. This means that a UV transmissive material must be used for the mold. Low-pressure, room-temperature processing means that UV-NIL has a strong potential for low-cost, high-volume production of nanostructured surfaces. Generally, if a hard mold is required, quartz is used and if a soft mold is needed, a polymer film is used. On the other hand, we have developed RT-NIL, which is direct nanoimprinting using sol-gel materials. RT-NIL does not require a resist thermal cycle and UV exposure when pressing a mold onto the resist. Therefore, RT-NIL offers us increased throughput and pattern accuracy in nanostructure fabrication. So far, we have used hydrogen silsesquioxane (HSQ) [3–20] and indium tin oxide (ITO) [21] as replicated sol-gel materials in RT-NIL.

HSQ is an inorganic sol-gel material consisting of repeating units of $H_2SiO_{3/2}$ [22]. HSQ is known to be a low-*k* dielectric [23, 24], and it is used as a high-resolution e-beam resist [25]. In addition, HSQ is used for optical applications because of its high transmittance in the range of visible light and good reflective index properties [26]. HSQ can also be the favorable etching mask because HSQ resin contains no organic groups [6, 12]. An HSQ polymer has two different structures, cage and ladder, as shown in Fig. 5.2 [22]. The caged HSQ and the ladder HSQ are supplied by Dow Corning Co. (FOX-16) and Tokyo Ohka Co. (OCD-12), respectively. The room-temperature reaction is as follows. First, HSQ reacts quickly with H_2O in the atmosphere. Depending on the amount of H_2O and catalyst, the silane bonds are hydrolyzed in HSQ to form silanol groups at room temperature, as shown Eq. 5.1.

$$\text{-Si-H + HO-H-} \rightarrow \text{-Si-OH} + H_2 \tag{5.1}$$

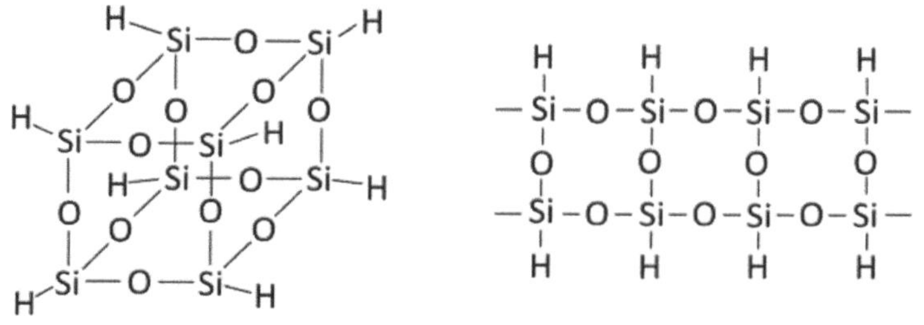

(a) cage structure (b) ladder structure

Figure 5.2 HSQ molecular structure. HSQ, hydrogen silsesquioxane.

Then the silanol bonds condense into siloxane bonds and H_2O molecules are produced as the by-product, as shown in Eq. 5.2.

$$-\text{Si-OH} + \text{HO-Si}- \rightarrow -\text{Si-O-Si}- + H_2O \quad (5.2)$$

As siloxane bonds are formed and H_2O molecules are displaced from the sol-gel skeletons, the density increases along with the increase in rigidity. There is a thermal stability difference between caged and ladder HSQ. During annealing, the ladder HSQ structure is stable but the caged HSQ structure is opened and transforms into a 3D network [27], which causes deformation of the caged HSQ pattern but not of ladder HSQ pattern after annealing.

We firstly used a SiO_2/Si hard mold for RT-NIL using spin-coated HSQ [3–13]. However, there were serious drawbacks, such as a higher imprinting pressure (10–40 MPa) and pattern size dependency of the residual layer thickness. To overcome the adobe problems, we developed RT-NIL using a combination of an HSQ-containing solvent (liquid-phase HSQ) and poly(dimethylsiloxane) (PDMS) mold to achieve a large-area nanoimprinting [15, 16]. PDMS is a silicone rubber with a highly hydrophobic surface property, high transparence, and high toughness [28–30]. In addition, PDMS has the characteristic of gas and vapor permeability [28]. Therefore, PDMS can be used as an NIL mold, because the solvent is appropriately chosen so that it is readily absorbed by the PDMS mold and evaporates into air by diffusion through the mold [31]. So far, the PDMS mold has been applied to T-NIL [32] and UV-NIL [33]. We first applied the PDMS mold to RT-NIL using HSQ resin.

5.2 RT-NIL Using a PDMS Mold

5.2.1 Patterning

We used two kinds of PDMS in this experiment [16]. One was the soft PDMS (s-PDMS; Dow Corning Co., Sylgard 184) and the other was the hard PDMS (h-PDMS; Shin-Etsu Chemical Co., X-32-3095). The PDMS parameters are listed in Table 5.1. The h-PDMS was 1.5 times harder than the s-PDMS. The s- or h-PDMS mold was fabricated by spin-coating PDMS onto the SiO_2/Si master mold, which was treated with a fluorinated antisticking agent (HARVES Co., DURASURF HD1101-TH). The rotating speed and time were 500 rpm and 3 s, respectively. Then, the s-PDMS-coated master mold was baked on a hot plate at 80°C for 30 min. On the other hand, the h-PDMS-coated master mold was baked on a hot plate at 150°C for 30 min. After curing of the PDMS, the PDMS mold was demolded from the SiO_2/Si mold. The thickness of s- or h-PDMS mold was about 2 mm.

Table 5.1 PDMS parameters

	s-PDMS	h-PDMS
Product name	Sylgard 184 (Dow Corning Co.)	X-32-3095 (Shin-Etsu Chemical Co.)
Hardness (Durometer Shore A)	51	78
Curing	80 °C, 30 min	150 °C, 30 min
Water contact angle (deg)	116	112

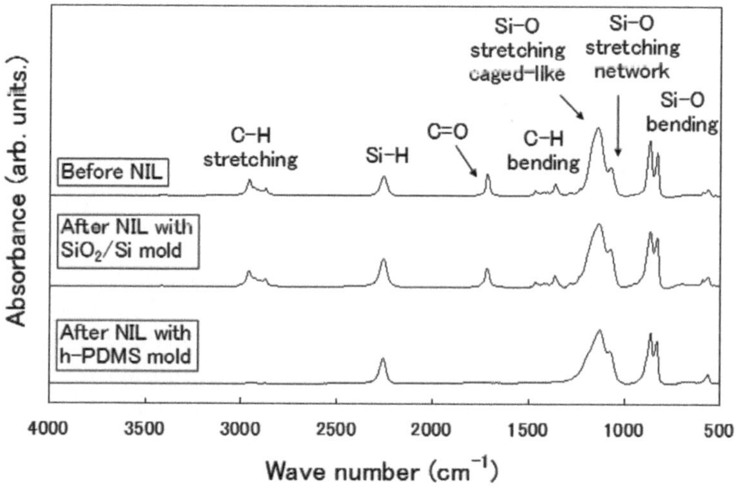

Figure 5.3 FTIR spectra of liquid-phase HSQ (FOX-16) on a Si substrate before and after NIL with a SiO_2/Si mold and an h-PDMS mold. FTIR, Fourier transform infrared; HSQ, hydrogen silsesquioxane.

Next, we confirmed whether the solvent evaporated after large-area imprinting, by using Fourier transform infrared (FTIR) spectroscopy [16]. Figure 5.3 shows the FTIR spectra of the liquid-phase HSQ (FOX-16) on the Si substrate before and after NIL with SiO_2/Si and an h-PDMS mold. The peaks for the liquid-phase HSQ film before NIL included C–H stretching at 2800–3000 cm^{-1}, Si–H stretching at 2260 cm^{-1}, C=O stretching at 1720 cm^{-1}, C–H bending at 1300–1500 cm^{-1}, Si–O stretching of the caged-like at 1130 cm^{-1} and network at 1070 cm^{-1}, and Si–O bending at 820–870 cm^{-1}. The organic peaks were derived from the solvent methylisobutylketone. After NIL with the SiO_2/Si mold, the organic peaks do not change. On the other hand, the organic peaks disappeared after NIL with the h-PDMS mold. This indicates that the h-PDMS mold could absorb the solvent completely.

Then, we optimized the RT-NIL process conditions (imprinting pressure and amount of FOX-16) using liquid-phase HSQ with a PDMS mold [16]. The conditions changed according to the shape and volumetric capacity of the pattern. We used 20 × 20 mm s- and h-PDMS molds having 1 μm line-and-space (L&S) patterns that were 500 nm in depth. Figure 5.4 shows the scanning electron microscopy (SEM) images of the imprinted HSQ pattern on a Si substrate using a PDMS mold. The transferability was improved when the amount of the FOX-16 liquid increased. Figure 5.4a shows the imprinted HSQ pattern depending on the amount of FOX-16 liquid using an h-PDMS mold at 1 MPa of imprint pressure. On the Si substrate, 0.5, 5, and 40 μL of FOX-16 were dropped. The HSQ patterns were successfully replicated using the larger amounts (5 and 40 μL) of FOX-16 liquid. However, the smaller amounts (0.5 μL) of the FOX-16 liquid could not fill up the cavity. Figure 5.4b shows the imprinted HSQ pattern depending on the imprinted pressure with s- and h-PDMS molds using 5 μL of the liquid. The pattern was successfully transferred at 0.2 MPa of imprint pressure using s- and h-PDMS molds. The 1 MPa of imprint pressure replicated the HSQ pattern with the h-PDMS mold, but the pattern got trapezoidally distorted with the s-PDMS mold because the pattern was crushed by the imprint pressure. Further, the pattern was completely crushed with more than 10 MPa of imprint pressure. The residual layer decreased as the imprint pressure increased. This indicates that the residual layer remained extremely thin when using 5 μL of FOX-16 liquid at 1 MPa with the h-PDMS mold. The fine HSQ pattern was replicated with s- and h-PDMS molds. However, the range of replicating the fine

HSQ pattern with the s-PDMS mold was narrow because the s-PDMS pattern was easily crushed by the imprint pressure.

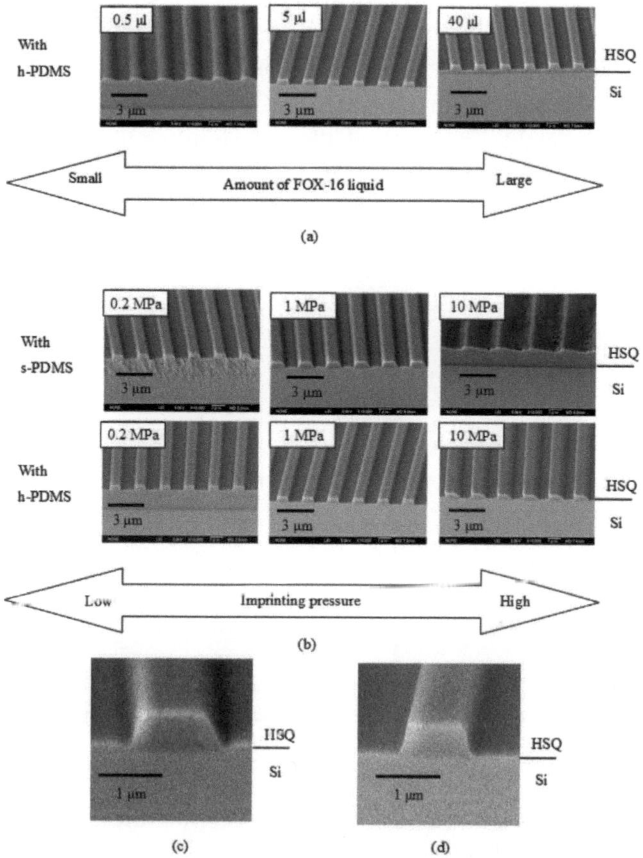

Figure 5.4 SEM images of an HSQ-replicated pattern with 1 μm L&S on a Si substrate using s- and h-PDMS molds. Transferability depends on (a) 0.5, 5, and 40 μL of FOX-16 liquid using an h-PDMS mold at 1 MPa and (b) imprint pressure of 0.2, 1, and 10 MPa with s- and h-PDMS molds using 5 μL of FOX-16 liquid. (c, d) 5 μL of FOX-16 liquid at 1 MPa with s- and h-PDMS molds, respectively. SEM, scanning electron microscopy; HSQ, hydrogen silsesquioxane; L&S, line-and-space; PDMS, poly(dimethyl siloxane).

It is expected that RT-NIL with an h-PDMS mold enables replication of large-area patterns. We carried out a large-area nanoimprinting with the h-PDMS mold. The 14 × 14 cm h-PDMS mold and a 4-inch Si wafer were imprinted by RT-NIL using liquid-phase HSQ (FOX-16) resin at 1 MPa of imprint pressure. The HSQ pattern was successfully replicated on the 4-inch Si wafer over the entire area [16]. The imprinting using liquid-phase HSQ provided various HSQ patterns with a line width ranging from 150 nm to 100 μm. Figure 5.5 shows an imprinted HSQ pattern 150 nm in width and 450 nm in height. The method allowed simultaneous imprinting of arbitrary patterns, including patterns of submicron sizes and those greater than 100 μm, in a single step by using an h-PDMS mold.

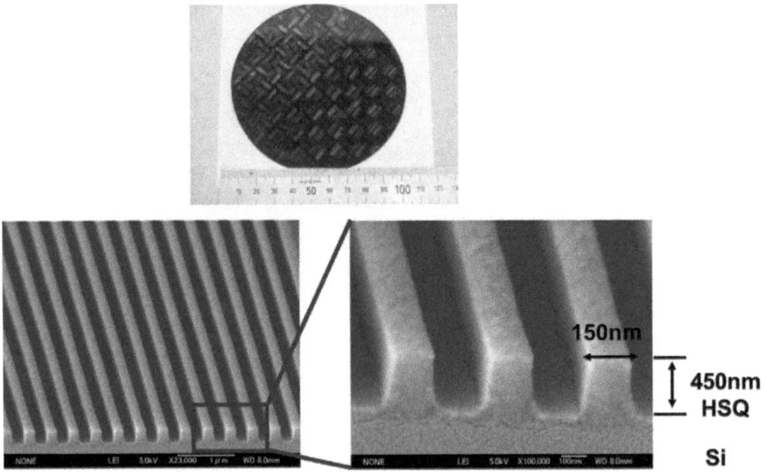

Figure 5.5 SEM image of an imprinted HSQ pattern 150 nm in width and 450 nm in height on a 4-inch Si wafer using FOX-16 and an h-PDMS mold. The imprint pressure was 1 MPa. SEM, scanning electron microscopy; HSQ, hydrogen silsesquioxane; PDMS, poly(dimethyl siloxane).

It is well known that the properties of thin films differ from those measured in bulk; therefore, there may be differences between the mechanical characteristics of an imprinted structure and a thin film. Hence, we have estimated Young's modulus of an imprinted ladder HSQ pillar from the spring constant, measured by using a Si cantilever manipulated with a three-axis actuator [34, 35]. The sizes of the HSQ pillars were 400, 600, and 1050 nm, and they were 3 μm in height. The spring constant determined Young's modulus of the imprinted pillar. The Young's modulus value of imprinted HSQ pillars and 4-μm-thick HSQ thin film were calculated as 4.6 ± 1.1

and 6.9 ± 0.3 GPa, respectively. Moreover, we measured the resonant frequency of the imprinted pillar by using the alternating current electrostatic force. Using the results of Young's modulus and the resonant frequency, we determined the density of the imprinted pillar. The density value of the imprinted pillar using HSQ was calculated as 1.08 ± 0.23 g/cm^3. The density of the resin fabricating the pillar structure was compared with that of the film. The density of the film made from HSQ was calculated as 1.68 ± 0.06 g/cm^3. On the other hand, the density of the imprinted pillar was lower than that of the film. Next, the atomic composition of the imprinted pillar and that of the film were examined using scanning electron microscopy–energy dispersive X-ray spectroscopy (SEM-EDX) (JEOL Ltd., EX-23000BU). The electron accelerating voltage was 5.0 kV. The atomic composition of the HSQ film was 42% silicon and 58% oxygen, and that of the HSQ pillar was 42% silicon and 58% oxygen. The atomic composition of the imprinted pillar was the same as that of the film. A possible reason for the lower density of the pillar structures compared with that of the thin films is the nanoporous structure of the PDMS mold. In the case of the pillar structures in the PDMS mold, the solvent evaporated through the porous structure. In the case of the film, however, the solvent evaporated into open air; it took more time to evaporate from the pillar mold than from the thin film. The ratio of solute to solvent in the pillar structure was, therefore, smaller than that in the thin film, since the solvent trapped in the imprinting material may not have had enough time to escape during the imprinting process. As the result, the density of the imprinted pillar was lower than that of the film.

5.2.2 Three-Dimensional Nanostructure Fabrication

Next, we carried out 3D nanostructure fabrication by RT-NIL. So far, LaFratta and coworkers have demonstrated that microtransfer molding (μTM) [36] could replicate a 3D structure created by multiphoton absorption polymerization [37]. PDMS can be used as a mold for 3D structures because it is elastic. μTM begins with the injection of a resist into the mold. Then, the mold is pressed onto a Si substrate and the resist is cured. Finally, the mold is removed to leave the replicated structure. We proposed a new 3D replication method [38] that uses a combination of focused ion beam–chemical vapor deposition (FIB-CVD) [39] and RT-NIL. The 3D master structure was fabricated by FIB-CVD. We used a commercially available FIB system (SIM2050SM2, SII Nano Technology Inc.) with a Ga^+ ion beam operating at 30 keV. FIB-CVD was carried out using a precursor of phenanthrene ($C_{14}H_{10}$) as the source material. The deposited material was diamond-like carbon (DLC), which was confirmed by Raman spectroscopy. Young's modulus was 187 GPa. DLC has sufficient hardness to be used as a mold. By combining the lateral growth mode with beam scanning, we obtained a 3D structure. The 3D computer-aided design (CAD) data and an SEM image are shown in Figs. 5.6a and 5.6b, respectively. The diameter was 2 μm, and the height was 3.2 μm. The rim thickness was 150 nm, and the stem width was 600 nm. We have successfully fabricated the master mold of the DLC wineglass. Next, we fabricated an s-PDMS copy mold by using the DLC master mold. The DLC master mold needed to be coated with a fluorinated antisticking agent to prevent adhesion. The thickness of the s-PDMS copy mold was about 2 mm. We used ladder HSQ. The imprinting pressure of 0.5 MPa was maintained for 5 min. at room temperature, following which the s-PDMS mold was removed from the Si substrate. As the result, the shape was deformed, but the 3D structure was found to be replicated on the substrate, as shown in Fig. 5.6c.

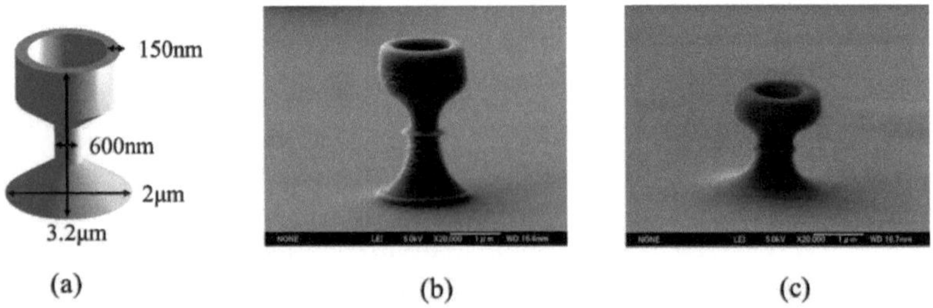

Figure 5.6 Microwineglass: (a) 3D CAD data and (b) SEM image of an FIB-CVD wineglass; (c) SEM image of an imprinted HSQ wineglass. The imprint pressure and time were 0.5 MPa and 5 min., respectively. CAD, computer-aided design; SEM, scanning electron microscopy; FIB-CVD, focused ion beam–chemical vapor deposition.

5.3 Example of NIL

5.3.1 Moth Eye Structure Fabrication

The phenomenon of antireflection is widely used to prevent insertion losses at the interfaces between different optical media. The moth eye pattern, with a tapered profile, a high aspect ratio, and a pitch smaller than the wavelength of the light, ensures that the effective refractive index near the interface changes gradually and reduces for a wide spectral bandwidth [40]. Many techniques are available for fabricating moth eye patterns [41–43]. NIL is a very useful technique to make nanopatterns using a simple process. So far, moth eye nanopatterns have been formed using T- and UV-NIL processes and increased the transmittance of the glass [44, 45]. However, moth eye–imprinted patterns using organic resist do not have high hardness or heat resistance. So, we fabricated Si*Ox* moth eye structures on a glass by RT-NIL using liquid-phase HSQ with h-PDMS [17]. This process replicated nanostructures at room temperature, low pressure, and high throughput. Moreover, the excellent property of HSQ after annealing was due to its chemical similarities to SiO_2.

Figures 5.7a and 5.7b show the imprinted patterns of caged and ladder HSQ before annealing, respectively. The imprint pressure and time were 1 MPa and 2 min., respectively. The pitch and height are 270 nm and 400 nm. Following this, the imprinted HSQ patterns were annealed at 600°C for 1 h in vacuum. Figure 5.7c shows that the imprinted pattern of caged HSQ deformed after 600°C annealing, as previously mentioned. On the other hand, the profile of the ladder HSQ pattern remained after 600°C annealing, as shown in Fig. 5.7d.

Figure 5.8 shows the result of the transmittance of glass with the imprinted structure before and after annealing. While the bare glass showed only 91.7% at 600 nm wavelength, the transmittance values of glass with caged HSQ patterns and ladder HSQ patterns before annealing increased to 94.4% and 95.3%, respectively. However, the transmittance value of glass with caged HSQ after annealing decreased drastically, to 92.6%, due to the degradation of the pattern profile and that with ladder HSQ after annealing decreased slightly, to 94.8%. The result indicates that ladder HSQ is superior to caged HSQ for making heat-resistant optical elements.

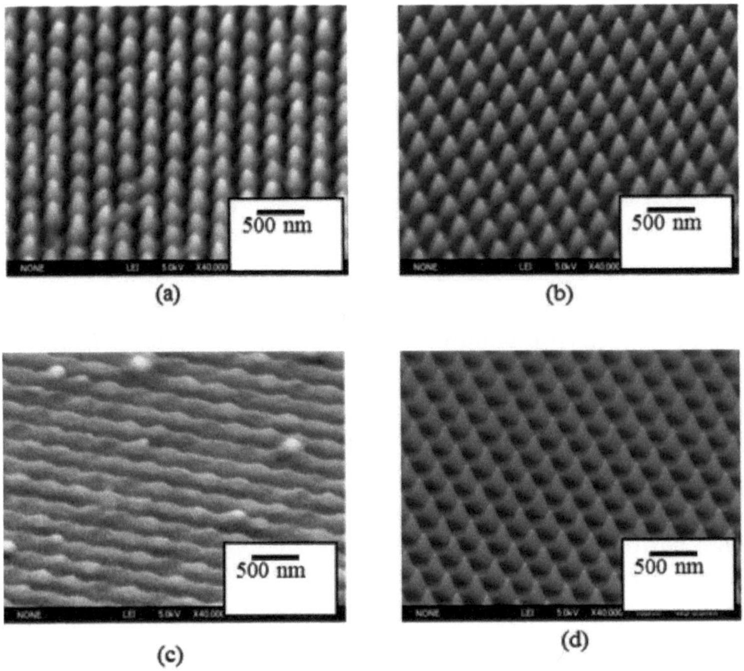

Figure 5.7 SEM images of imprinted patterns. The imprint pressure and time were 1 MPa and 2 min., respectively: (a) caged HSQ before annealing, (b) ladder HSQ before annealing, (c) caged HSQ after 600°C annealing, and (d) ladder HSQ after 600°C annealing. SEM, scanning electron microscopy; HSQ, hydrogen silsesquioxane.

5.3.2 Gold Nanoparticle Formation Using an Imprinted HSQ Pattern for SERS

The fabrication of gold nanoparticle (Au NP) linear assemblies was carried out using the imprinted HSQ pattern [46]. HSQ is an inorganic material and so is suitable as a surface-enhanced Raman spectroscopy (SERS) substrate for the organic chemical material. We revealed that the SERS substrate fabricated via HSQ consisting of inorganic polymer was suitable for organic chemical analysis by comparing that fabricated via SU-8 consisting of organic polymer. In this experiment, ladder HSQ was used because imprinted HSQ was annealed at 600°C to transform into SiO_x.

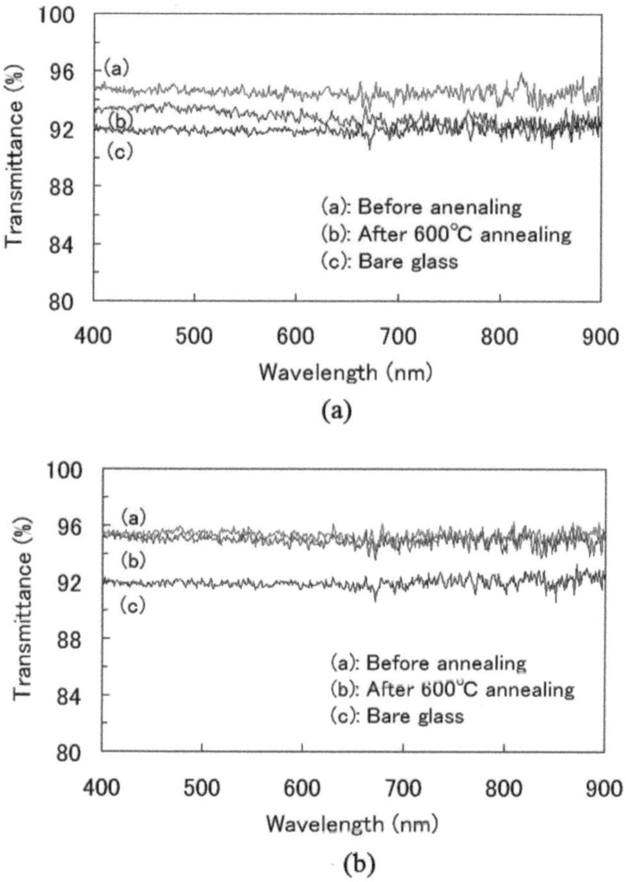

Figure 5.8 Transmittance of glass with an imprinted structure before and after annealing at 600°C: (a) caged HSQ and (b) ladder HSQ. HSQ, hydrogen silsesquioxane.

Raman spectroscopy was acquired using a near-infrared Raman spectrometer (RAM-100S, Lambda Vision Inc.) at an incident laser wavelength of 785 nm and power of 30 mW. On the substrate, 30 μL of 1 mM 4,4'-bipyridine (4bpy) aqueous solution was dropped, and then the measurement was immediately performed. Figure 5.9a shows the SERS spectra of 4bpy using the imprinted HSQ patterns with and without Au NPs of 100 nm diameter. The peaks of 4bpy are known as ring breathing at 1000 cm^{-1}, C–H in-plane bending at 1200 cm^{-1}, inter-ring stretching at 1280 cm^{-1}, and ring stretching at

1600 cm^{-1}. The characteristic enhanced Raman spectrum of 4bpy was observed with Au NPs. Next, the imprinted SU-8 patterns were applied to the SERS substrate. Figure 5.9b shows the SERS spectra of 4bpy using the imprinted SU-8 patterns with and without Au NPs. The peaks of C–O–C stretching at 900 cm^{-1}, epoxy stretching at 1100 cm^{-1}, C–H in-plane bending at 1200 cm^{-1}, inter-ring stretching at 1280 cm^{-1}, and ring stretching at 1600 cm^{-1} were not avoidable using the SU-8-based substrate. Thus, it is difficult to determine the 4bpy peaks due to the organic signals of the SU-8 substrate. In this way, the result demonstrated that the imprinted HSQ patterns with Au NPs were useful for SERS measurement.

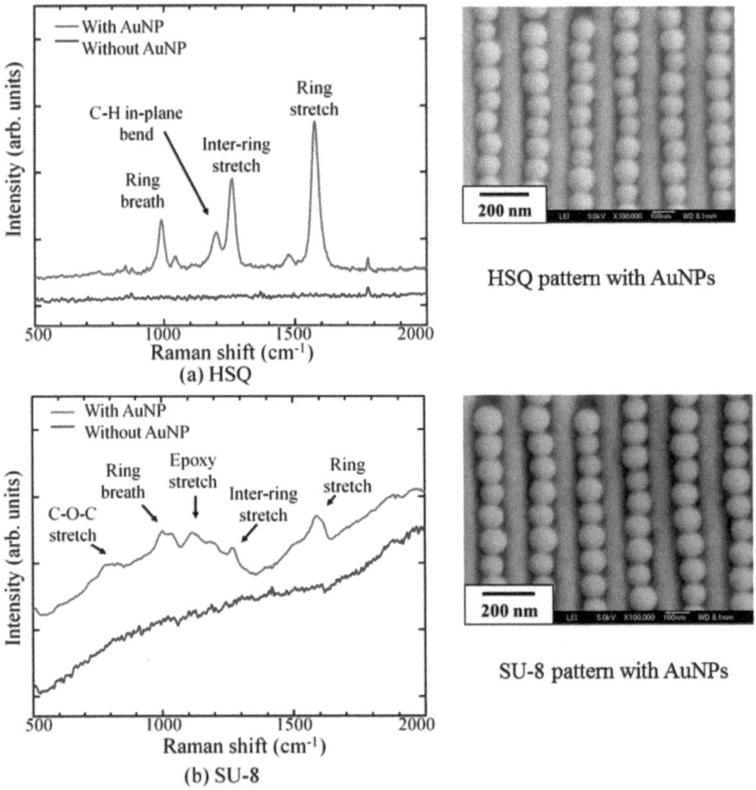

Figure 5.9 SERS spectra of 4bpy using Au NP assemblies on the imprinted patterns: (a) HSQ and (b) SU-8. SERS, surface-enhanced Raman spectroscopy; HSQ, hydrogen silsesquioxane.

5.4 Conclusions

We have developed RT-NIL using HSQ of both cage and ladder structures as a replicated material, which is direct nanoimprinting without using thermal cycle and UV exposure, resulting in an increase in the throughput and pattern accuracy. In the case of the spin-coating method using a hard mold such as SiO_2/Si, higher imprinting pressure is required and the imprinted HSQ depth depends on the line width. To overcome the above issues, a casting method using an s-PDMS mold has been developed, which successfully achieved HSQ pattern replication on a 4-inch Si wafer over the entire area with low pressure. Furthermore, 3D nanostructure, moth eye fabrication, and Au NP formation for SERS have been demonstrated. RT-NIL using ladder HSQ will be applied to make various devices, such as heat-resistant optical elements, because ladder HSQ transforms into a Si*Ox* structure without shape change after annealing at 600°C.

References

1. S. Y. Chou, P. R. Krauss, and P. J. Renstrom (1995). Imprint of sub-25 nm vias and trenches in polymers, *Appl. Phys. Lett.*, **67**, 3114–3116.
2. J. Haisma, M. Verheijen, K. van den Heuvel, and J. van den Berg (1996). Mold-assisted nanolithography: a process for reliable pattern replication, *J. Vac. Sci. Technol.*, **B14**, 4124–4128.
3. S. Matsui, Y. Igaku, H. Ishigaki, J. Fujita, M. Ishida, Y. Ochiai, M. Komuro, and H. Hiroshima (2001). Room temperature replication in spin on glass by nanoimprint technology, *J. Vac. Sci. Technol.*, **B19**, 2801–2805.
4. Y. Igaku, S. Matsui, H. Matsui, J. Fujita, M. Ishida, Y. Ochiai, H. Namatsu, M. Komuro, and H. Hiroshima (2002). Room temperature nanoimprint technology using hydrogen silsequioxane (HSQ), *Jpn. J. Appl., Phys.*, **41**, 4198–4202.
5. S. Matsui, Y. Igaku, H. Ishigaki, J. Fujita, M. Ishida, Y. Ochiai, H. Namatsui, and M. Komuro (2003). Room-temperature nanoimprint and nanotransfer printing using hydrogen silsequioxane, *J. Vac. Sci. Technol.*, **B21**, 688–692.
6. K. Nakamatsu, K. Watanabe, K. Tone, T. Katase, W. Hattori, Y. Ochiai, T. Matsuo, M. Sasago, H. Namatsu, M. Komuro, and S. Matsui (2004). Bi-layer resist method for room-temperature nanoimprint lithography, *Jpn. J. Appl. Phys.*, **43**, 4050–4053.

7. K. Nakamatsu, K. Watanabe, K. Tone, H. Namatsu, and S. Matsui (2005). Nanoimprint and nanocontact technologies using hydrogen silsesquioxane, *J. Vac. Sci. Technol.*, **B23**, 507–512.

8. K. Nakamatsu, K. Tone, H. Namatsu, and S. Matsui (2006). Room-temperature nanocontact printing using soft template, *J. Vac. Sci. Technol.*, **B24**, 195–199.

9. K. Nakamatsui and S. Matsui (2006). Nanoimprinting using liquid-phase hydrogen silsesquioxane, *Jpn. J. Appl. Phys.*, **45**, L546–L548.

10. M. Kawamori, K. Nakamatsu, Y. Haruyama, and S. Matsui (2006). Effect of oxygen plasma irradiation on hydrogen silsesquioxane nanopatterns replicated by room-temperature nanoimprinting, *Jpn. J. Appl. Phys.*, **45**, 8994–8996.

11. K. Nakamatsu, K. Ishikawa, N. Taneichi, and S. Matsui (2007). Room-temperature nanoimprinting using ladder hydrogen silsesquioxane, *Jpn. J. Appl. Phys.*, **46**, 5388–5390.

12. K. Nakamatsu, K. Ishikawa, N. Taneichi, and S. Matsui (2007). Room-temperature nanoimprinting using ladder hydrogen silsesquioxane(HSQ), *J. Photopolym. Sci. Technol.*, **20**, 555–558.

13. K. Nakamatsu, M. Okada, and S. Matsui (2008). Fabrication of high-aspect Si structures by deep reactive ion etching using hydrogen silsesquioxane masks replicated by room temperature nanoimprinting, *Jpn. J. Appl. Phys.*, **47**, 8619–8621.

14. Y. Kang, M. Okada, K. Nakamatsu, Y. Haruyama, K. Kanda, and S. Matsui (2009). Room-temperature nanoimprint using sol-gel ITO film, *J. Photopolym. Sci. Technol.*, **22**, 189–192.

15. Y. Kang, M. Okada, K. Nakamatsu, Y. Haruyama, K. Kanda, and S. Matsui (2009). Room-temperature nanoimprint using liquid-phase hydrogen silsesquioxane with PDMS mold, *J. Photopolym. Sci. Technol.*, **22**, 193–194.

16. Y. Kang, M. Okada, C. Minari, K. Kanda, Y. Haruyama, and S. Matsui (2010). Room-temperature nanoimprinting using liquid-phase hydrogen silsesquioxane with hard poly(dimethylsiloxane) mold, Jpn. *J. Appl. Phys.*, **49**, 06GL13.

17. Y. Kang, M. Okada, K. Kanda, Y. Haruyama, and S. Matsui (2011). SiOx moth-eye structure fabrication by room-temperature nanoimprint using hydrogen silsesquioxane, *Proc. RadTech Asia, Jpn.*, 36–39.

18. K. Kang, S. Omoto, Y. Nakai, M. Okada, K. Kanda, Y. Haruyama, and S. Matsui (2011). Nanoimprint replication of nonplanar nanostructure fabricated by focused-ion-beam chemical vapor deposition, *J. Vac. Sci. Technol.*, **B29**, 011005.

19. K. Kang, M. Okada, S. Omoto, Y. Haruyama, K. Kanda, and S. Matsui (2011). Room temperature nanoimprinting using spin-coated hydrogen silsesquioxane with high boiling point solvent, *J. Vac. Sci. Technol.*, **B29**, 06FC03.

20. Y. Kang, Y. Nakai, Y. Haruyama, and S. Matsui (2012). Density measurement of pillar structure fabricated via nanoimprinting a poly(dimethylsiloxane) mold, *J. Vac. Sci. Technol.*, **B30**, 06FB06.

21. Y. Kang, M. Okada, K. Nakamatsu, K. Kanda, Y. Haruyama, and S. Matsui (2009). UV irradiation effect on sol-gel indium tin oxide nanopatterns replicated by room-temperature nanoimprint, *J. Vac. Sci. Technol.*, **B27**, 2805–2809.

22. J. Zhao, I. Malik, T. Ryan, E. T. Ogawa, P. S. Ho, W. Shih, A. J. Mckerrow, and K. J. Taylor (1999). Thermomechanical properties and moisture uptake characteristics of hydrogen silsesquioxane submicron films, *Appl. Phys. Lett.*, **74**, 944–946.

23. C. L. Frye and W. T. Collins (1970). Oligomeric silsesquioxanes, (HSiO3/2)n, *J. Am. Chem. Soc.*, **92**, 5586–5588.

24. J. H. Wang, W. J. Chen, T. C. Chang, P. L. Liu, S. L. Cheng, J. Y. Lin and L. J. Chen (2003). Structural and electrical characteristics of low-dielectric constant porous hydrogen silsesquioxane for Cu metallization, *J. Electrochem. Soc.*, **150**, F141–F146.

25. H. Namatsu, Y. Takahashi, K. Yamazaki, T. Yamaguchi, M. Nagase, and K. Kurihara (1998). Three-dimensional siloxane resist for the formation of nanopatterns with minimum linewidth fluctuations, *J. Vac. Sci. Technol.*, **B16**, 69–76.

26. S. L. Neale, M. P. Macdonald, K. Dholakia, and T. F. Krauss (2005). All-optical control of microfluidic components using form birefringence, *Nat. Mater.*, **4**, 530–533.

27. Y. K. Siew, G. Sarkar, X. Hu, J. Hui, A. See and C. T. Chua (2000). Thermal curing of hydrogen silsesquioxane, *J. Electrochem. Soc.*, **147**, 335–339.

28. Y. Xia and G. M. Whitesides (1998). Soft lithography, *Angew. Chem., Int. Ed.*, **37**, 550–575.

29. K. Efimenko, W. E. Wallance and J. Genzer (2002). Surface modification of sylgard-184 poly(dimethyl siloxane) networks by ultraviolet and ultraviolet/ozone treatment, *J. Colloid. Interface Sci.*, **254**, 306–315.

30. M. J. Lee, N. Y. Lee, J. R. Lim, J. B. Kim, M. Kim, H. K. Baik, and Y. S. Kim (2006). Antiadhesion surface treatments of molds for high-resolution unconventional lithography, *Adv. Mater.*, **18**, 3115–3119.

31. E. Kim, Y. Xia, X.-H. Zhao, and G. M. Whitesides (1997). Solvent-assisted microcontact molding: a convenient method for fabricating three-dimensional structures on surfaces of polymers, *Adv. Mater.*, **9**, 651–654.

32. K. Yang, K. Yoon, K. Choi, and H. Lee (2009). The direct nano-patterning of ZnO using nanoimprint lithography with ZnO-sol and thermal annealing, *Microelectron. Eng.*, **86**, 2228–2231.

33. K. Y. Suh, Y. S. Kim, and H. H. Lee (2001). Capillary force lithography, *Adv. Mater.*, **13**, 1386–1389.

34. Y. Kang, M. Okada, Y. Nakai, Y. Haruyama, K. Kanda, and S. Matsui (2011). Mechanical characteristics of imprinted nanostructures fabricated with a poly(dimethylsiloxane) mold, *J. Vac. Sci. Technol.*, **B29**, 06FC10.

35. Y. Kang, Y. Nakai, Y. Haruyama, and S. Matsui (2012). Density measurement of pillar structure fabricated via nanoimprinting a poly(dimethylsiloxane) mold, *J. Vac. Sci. Technol.*, **B30**, 06FB06.

36. C. N. LaFratta, T. Baldacchini, R. A. Farrer, J. T. Fourkas, M. C. Teich, B. E. A. Saleh, and M. J. Naughton (2004). Replication of two-photon-polymerized structures with extremely high aspect ratios and large overhangs, *J. Phys. Chem.*, **B108**, 11256–11258.

37. S. Kawata, H. Sun, T. Tanaka, and K. Takada (2001). Finer features for functional microdevices, *Nature*, **412**, 697–698.

38. K. Kang, S. Omoto, Y. Nakai, M. Okada, K. Kanda, and S. Matsui (2011). Nanoimprint replication of nonplanar nanostructure fabricated by focused-ion-beam chemical vapor deposition, *J. Vac. Sci. Technol.*, **B29**, 011005.

39. S. Matsui, T. Kaito, J. Fujita, M. Komuro, K. Kanda, and Y. Haruyama (2000). Three-dimensional nanostructure fabrication by focused-ion-beam chemical vapor deposition, *J. Vac. Sci. Technol.*, **B18**, 3181–3184.

40. P. B. Clapham and M. C. Hutley (1973). Reduction of Lens reflexion by the "moth eye" principle, *Nature*, **244**, 281–282.

41. C.-H. Sun, P. Jiang, and B. Jiang (2008). Broadband moth-eye antireflection coatings on silicon, *Appl. Phys. Lett.*, **92**, 061112–06115.

42. W. Min, A. P. Betancourt, P. Jiang, and B. Jiang (2008). Bioinspired broadband antireflection coatings on GaSb, *Appl. Phys. Lett.*, **92**, 141109–141114.

43. H. Toyota, K. Tanaka, M. Okano, T. Yotsuya, and H. Kikuta (2001). Fabrication of microcone array for antireflection structured surface using metal dotted pattern, *Jpn. J. Appl. Phys.*, **40**, L747–L749.

44. K. Han, H. Lee, D. Kim, and H. Lee (2009). Fabrication of anti-reflection structure on protective layer of solar cells by hot-embossing method, *Sol. Energy. Mater. Sol. Cells.*, **93**, 1214–1217.

45. K. Han, J. Shin, and H. Lee (2010). Enhanced transmittance of glass plates for solar cells using nano-imprint lithography, *Sol. Energy. Mater. Sol. Cells*, **94**, 583–587.

46. Y. Kang, T. Fukuoka, R. Takahashi, Y. Utsumi, Y. Haruyama, and S. Matsui (2014). Fabrication of gold nanoparticle pattern using imprinted hydrogen silsesquioxane pattern for surface-enhanced Raman scattering, *Microsyst. Technol.*, **20**, 1993–2000.

Chapter 6

Basic Mechanisms of Nanoimprint Lithography

Yoshihiko Hirai
Department of Physics and Electronics Engineering,
Graduate School of Engineering, Osaka Prefecture University,
Gakuen-cho, Sakai, 599-8531 Osaka, Japan
hirai@pe.osakafu-u.ac.jp

6.1 Introduction

Thermal nanoimprint lithography (T-NIL) [1] is applicable for various kinds of materials, such as thermos plastic resin, organic semiconductors, glass, and other functional materials. In the T-NIL process, a mold having fine structures is pressed to the thermoplastic material above its glass transition temperature and the fine structures are transferred to the resin. The process is very simple; however, in the early days [2–4], some issues were reported relative to the resin deformation process. These are important issues to avoid defect problems when designing optimum process conditions. Understanding the mechanism of the NIL process is a key issue to designing the NIL process and eliminating fatal defects.

Nanoimprint and Its Applications
Edited by Akihiro Miyauchi

ISBN 978-981-4800-37-2 (Hardcover), 978-0-429-03192-2 (eBook)
www.jennystanford.com

In this chapter, resin deformation processes are investigated by simulation and experimental studies for various aspect ratios of the mold patterns and various initial resin thicknesses [5]. On the basis of the results, we discuss process optimization for high-aspect-ratio pattern fabrication by T-NIL [6]. In addition, the time-dependent deformation process is investigated and we discuss process conditions [7, 8].

6.2 Basic Process Mechanisms [5]

In this section, numerical simulation and experimental studies are described to understand basic mechanism of the T-NIL process.

In the general T-NIL process, a thermoplastic resin is heated above its glass transition temperature to soften the resin. Over the glass transition temperature, the resin is assumed to be a rubber elastic body [9, 10], which can adapt to a large deformation. The characteristics of rubber is expressed by the Mooney–Rivlin model [11, 12].

The stress S_{ij} is expressed as follows:

$$S_{ij} = \frac{\partial W}{\partial \varepsilon_{ij}}, \tag{6.1}$$

where W is a strain density function and ε_{ij} is strain. The W is expressed as follows:

$$W = C_{10}(I_1 - 3) + C_{01}(I_2 - 3) \tag{6.2}$$

Here,

$$I_1 = \lambda_1^2 + \lambda_2^2 + \lambda_3^2, \tag{6.3}$$

$$I_2 = \lambda_1^2\lambda_2^2 + \lambda_2^2\lambda_3^2 + \lambda_3^2\lambda_1^2, \tag{6.4}$$

and the C_{01} and C_{10} are Moony constants and the λ is is expansion rate. These constants are obtained from the following approximations [13]:

$$C_{01} = 0.25C_{10} \tag{6.5}$$

$$6(C_{10} + C_{01}) \approx E, \tag{6.6}$$

where E is an elastic modulus of the resin at the process temperature.

In this work, we use the commercially available software MSC-Marc, which is suitable for large deformations in materials under contact boundary conditions using automatic variable mesh processing. In this modeling, any resin chain interaction, such as slipping or ordering, is ignored. Also, air trapping, bubbling, or absorption into the resin during press is out of consideration because the applied pressure for resin is very high compared with the air ambient pressure and it might have some effect on the resin deformation macroscopically.

Figure 6.1a shows the schematic diagram of the analysis model. The mold has a periodic line-and-space (L&S) pattern. The line width L is the groove width of the mold pattern, and the space S is the distance between the grooves. The depth of the groove is h, and the initial thickness of the resin is t. The aspect ratio A and the duty ratio D are defined as follows:

$$A = h/L \tag{6.7}$$

$$D = L/(L + S) \tag{6.8}$$

The aspect ratio A and the normalized initial resin thickness t/h are modulated, while the groove depth h is maintained at unity.

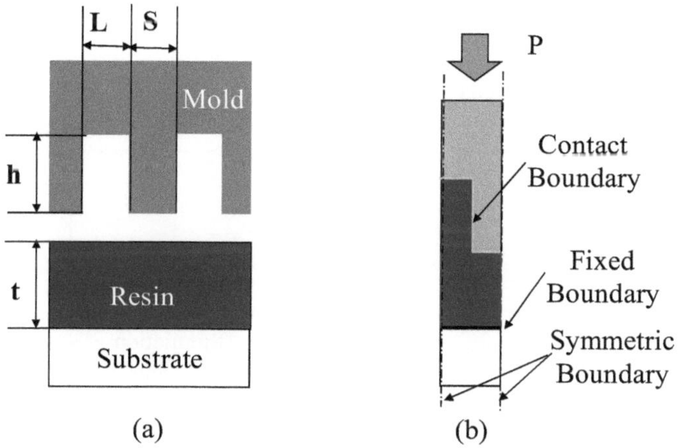

Figure 6.1 Schematic diagram of the analysis system: (a) Definition of the pattern configuration. (b) Simulation system and conditions.

Considering the symmetry of the system, we simulate the periodic part. The boundary conditions and material definitions are shown in Fig. 6.1b.

6.3 Result and Discussion

6.3.1 Impact of the Applied Pressure

Firstly, the deformation process is discussed. The filling rate of the resin, which is defined as the ratio of the resin area filled into the groove to the cross section of the groove, is investigated in terms of the applied pressure. Figure 6.2 shows the relations between the filling ratio and the relative applied pressure P/P_0, where P_0 is the required pressure for complete deformation. The dashed line shows the simulation result when the aspect ratio is 1.6, and the dots show related experimental results. As the applied pressure increases, the resin is filled into the groove. However, the filling ratio is not proportional to the applied pressure. Figure 6.3 shows the cross-section profiles of the resins in experiments and simulations for various applied pressures. Those results fairly agree with each other. In the simulation, the resin is almost filled into the groove (i.e., a filling rate of above 90%) when the relative pressure is around 60%. This indicates that perfect deformation needs large pressure. However, lower pressure is allowed for moderate deformation of the resin in the NIL process.

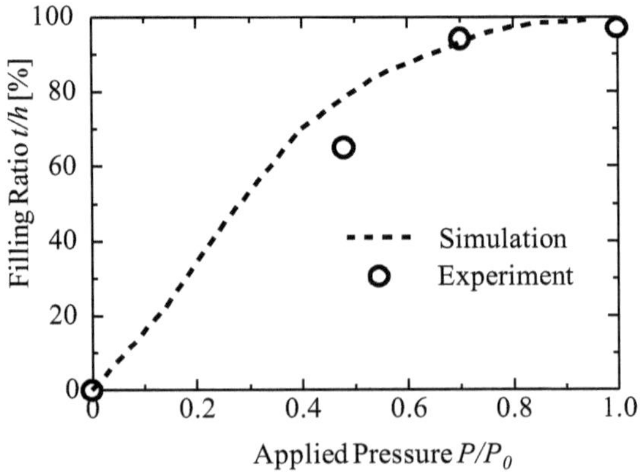

Figure 6.2 Relations between related applied pressures and filling coefficients. Dashed line: simulation. Circles: experiments (PMMA on Si, 150°C, P_0 = 20 MPa, Mw = 996 k).

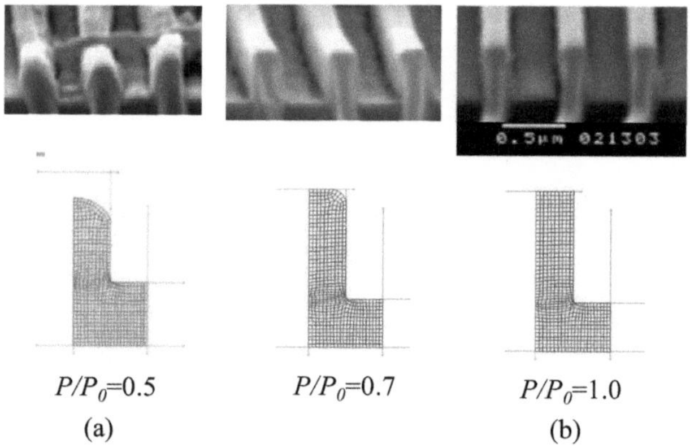

Figure 6.3 Cross-sectional profiles for various applied pressures: (a) Experimental results and (b) simulation results.

6.3.2 Impact of the Aspect Ratio

When the line width becomes narrow, the required relative pressure may increase because the resin needs large deformation to fill into the cavity. We evaluate the required relative pressure P/E to fill the resin into the mold grooves, where P is the pressure applied to the back side of the mold and E is elastic modulus of the resin at process temperature.

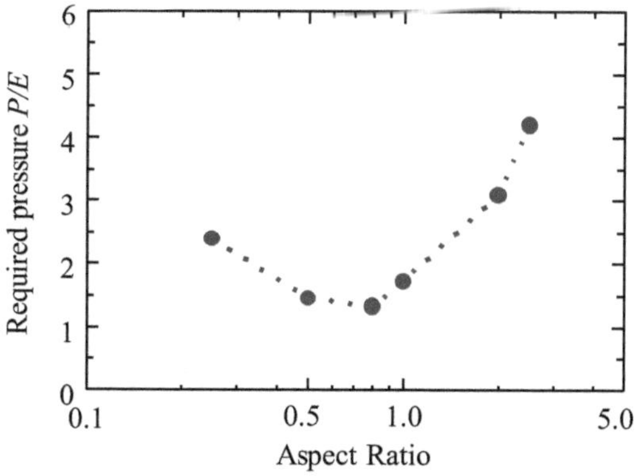

Figure 6.4 Relation between the aspect ratio of the pattern and the required relative pressure for fully filling the mold cavity (t/h = 1).

Figure 6.4 shows the simulation result of the required relative pressure P/E for resin filling into the mold cavity in various aspect ratios A when the initial thickness of the resin (t/h) is 1. When the aspect ratio is above around 0.8, the required pressure increases because the resin needs large deformation to fill into the narrower cavity.

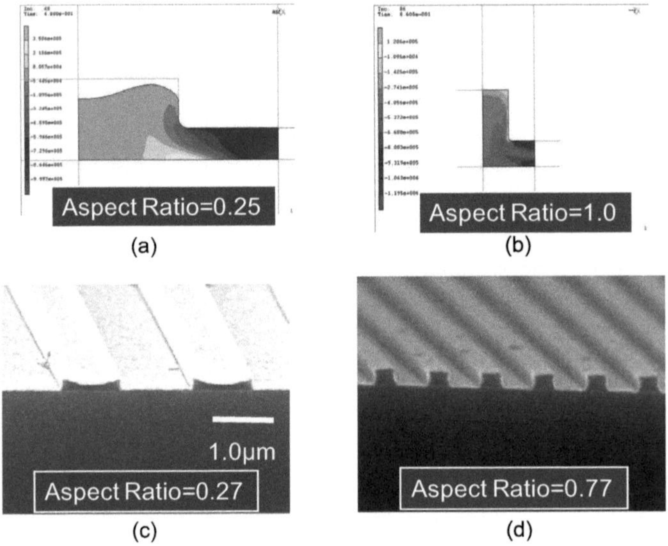

Figure 6.5 Cross-sectional profiles of the deformed resin for various aspect ratios under the same pressure condition. (a, b) Simulations (P/E = 1.8) and deformed shape and stress distribution. (c, d) Experiments (PMMA on Si, 170°C, P_0 = 30 MPa, Mw = 660 k).

On the contrary, higher pressure is also required for lower-aspect-ratio patterns (an aspect ratio below 0.8). Figures 6.5a and 6.5b show the simulated cross-sectional profiles, where the aspect ratios of the patterns are 0.25 and 1, respectively. In these cases, the applied pressure is the same. When the aspect ratio is 0.25, the resin is deformed only near the edge portion of the cavity and the center portion of the cavity is not filled enough. Additional pressure is required to completely fill the resin into the cavity. Figures 6.5c and 6.5d show the related experimental results, where the aspect ratios of the pattern cavity are 0.27 and 0.77, respectively. The experimental results show that the resin is completely filled into the cavity at a large aspect ratio (A = 0.77), while only edge portion of the resin is deformed at a low aspect ratio (A = 0.27). The simulation and the related experimental results agree well with each other. These results show that not only high-aspect-ratio patterns but also low-aspect-ratio patterns, such as a contact pad or an electrode pattern, require large pressure. The required pressure is minimized when the aspect ratio of the groove is around 0.8.

6.3.3 Impact of the Initial Resin Thickness

The impact of the initial thickness of the resin is considered. Figure 6.6 shows the relations between the required relative pressure (P/E) for complete filling as a function of the initial relative thickness of the resin (t/h) for various aspect ratios. When the initial relative thickness decreases to less than around t/h = 2, the required pressure increases. Figures 6.7a and 6.7b show the cross-section profiles of the experimental and simulations results under the same pressure. Figures 6.7c and 6.7d illustrate the deformation mechanism. When the initial thickness decreases, the resin is not sufficiently filled into the cavity pattern. This is because the resin close to the substrate is fixed on the substrate and hardly deforms laterally (sheared) and the flow into the cavity is restricted in thin films. When the initial thickness increases, the influence from the fixed boundary is relaxed and the resin flows easily into the pattern cavity.

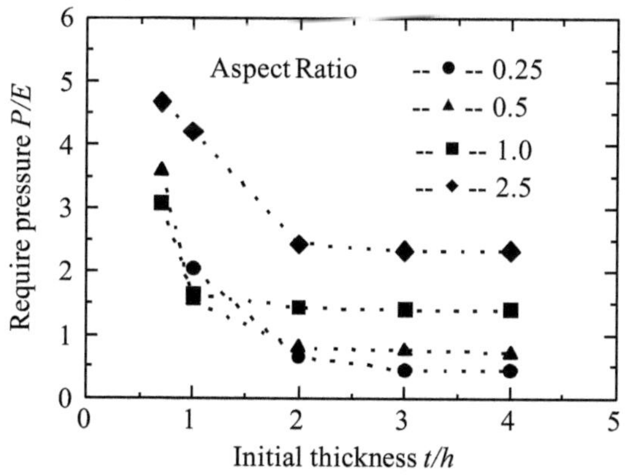

Figure 6.6 Relations between initial resin thickness and required pressure for various aspect ratio patterns.

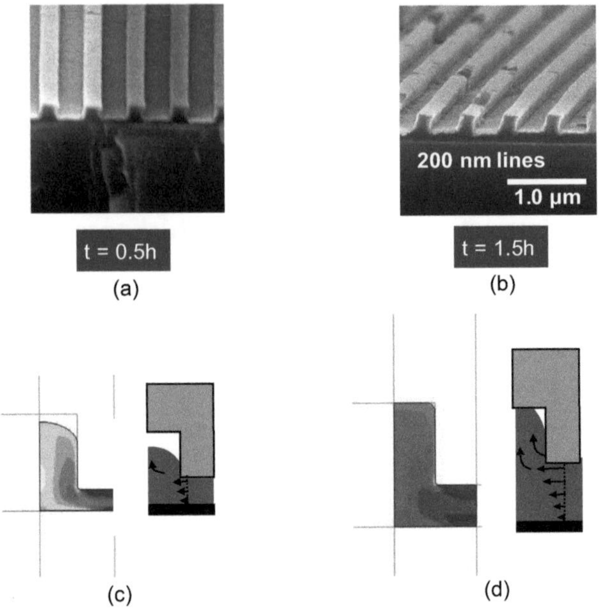

Figure 6.7 Cross-sectional profiles of the resins in a medium-aspect-ratio pattern for various initial resin thicknesses. (a, b) Experimental results and (c, d) simulation result and schematics of the resin deformation process.

This mechanism causes a different deformation process. Figures 6.8a and 6.8b show the experimental cross-section photos of the resin for t/h = 1 and 2, respectively. The relative applied pressure (P/P_0) is 0.5, where P_0 is the required pressure for the resin to completely fill into the cavity. The cross sections of the resin differ from each other. When the initial thickness is less, the resin is bunched up at the edge of the mold cavity. On the other hand, the resin rises up in the middle of the cavity pattern when the relative initial thickness becomes more.

Figures 6.8c and 6.8d show simulation results. The results agree with the experimental results. The reason is also illustrated in the figures. When the initial thickness is less, the resin is hardly moved along the lateral direction and the resin at the center is punched up. On the other hand, the resin moves easily when the thickness increases because the resin deforms easily and the center area of the resin is raised above the surroundings.

As discussed above, the initial resin thickness is strongly related to the resin deformation process and required pressure. The simulation results fairly agree with experimental results as per a rubber elastic model and simple approximations as discussed in Section 6.2 and help understand the deformation mechanism.

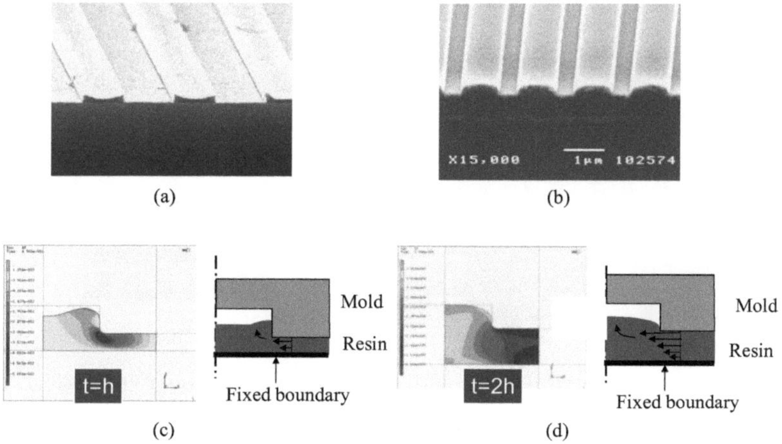

Figure 6.8 Cross-sectional profiles of the resins in a low-aspect-ratio pattern for various initial resin thicknesses. (a, b) Experimental results and (c, d) simulation results and schematic of the resin deformation process.

6.4 Defect Analysis and Process Optimization [6]

In this section, typical defects due to induced strain in the thermal nanoimprinting process are discussed. The typical case of defects in the deformation process arises in high-aspect-ratio pattern processing. We discuss the incidence of defects and how to deal with the defects on the basis of simulation works.

6.4.1 Typical Defect in High-Aspect-Ratio Pattern Fabrication

Figure 6.9a shows a typical defect after mold releasing in T-NIL. The resin is fractured and removed, as shown in the circled area. Figure 6.9b shows a schematic view of the defect. It might be fractured from the base of the pattern during the mold releasing process. This kind of defect becomes serious in a high-aspect-ratio pattern fabrication. We investigate the factors of this defect on the basis of computational work and optimize the process.

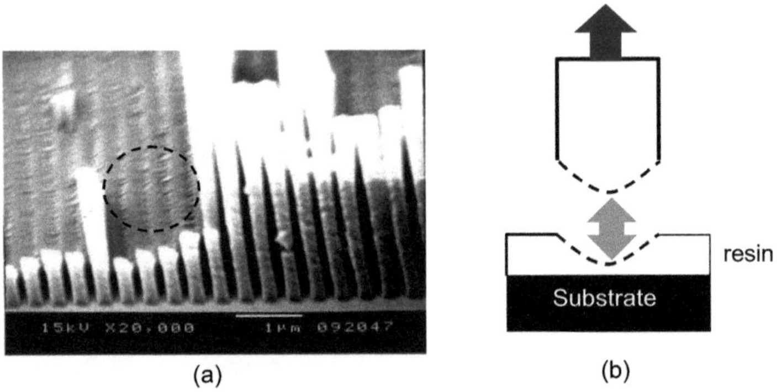

Figure 6.9 A typical fracture defect in thermal nanoimprint lithography. (a) SEM image of the fracture defect. (b) Schematic view of the fracture defect. SEM, scanning electron microscopy.

6.4.2 Simulation and Experiments

The process sequence is one of the important factors. Figure 6.10 shows a typical thermal nanoimprinting process. The first step is a pressing-and-holding process above the glass temperature of the resin (Fig. 6.10a). During this step, the resin turns soft and becomes deformed. After deformation is completed, the cooling process starts (Fig. 6.10b). Finally, the mold is released from the resin (Fig. 6.10c). Figure 6.10d shows typical process temperature and pressure diagrams. Figure 6.10e shows typical experimental results using the process sequence. The resin was polymethylmethacrylate (PMMA), and the molecular weight was 350 k. The thickness of the resin was 1 μm. The imprinting temperature was 170°C, and the release temperature 40°C. The imprinting pressure was 50 MPa, and the mold was held for 5 min. The line width varied from 75 nm to 1 μm, whereas the pattern height was 1 μm. Almost all the patterns are removed after the mold is released. In this case, the imprinting pressure is continuously applied to the mold until the mold is released, as shown in the process sequence in Fig. 6.10d. Also, the mold and the substrate are rapidly cooled down to 40°C using a water-cooling system for 5 min. In this case, the resin was damaged by the applied pressure.

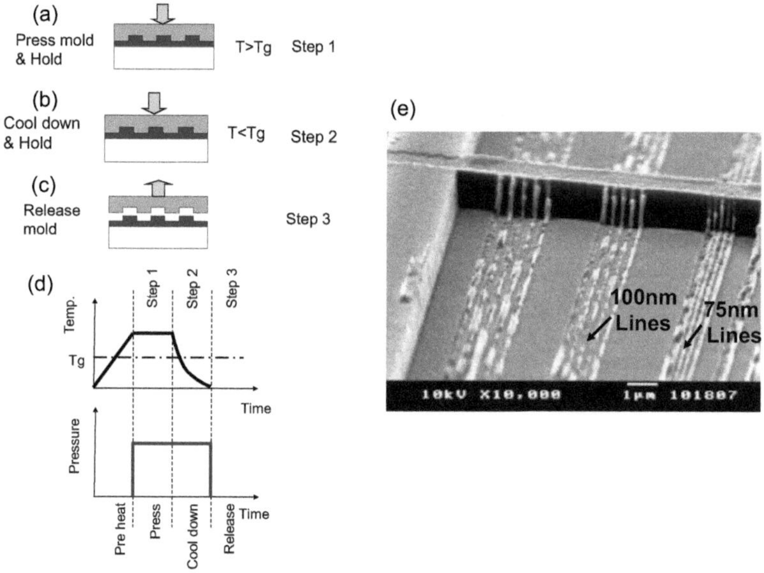

Figure 6.10 Typical process in thermal nanoimprint lithography. (a) Step 1: Pressing and holding. (b) Step 2: Cooling. (c) Step 3: Releasing. (d) Pressure and temperature sequence. (e) Experimental result.

To confirm the reason for the defect, stress distributions in each step were simulated.

6.4.2.1 Step 1: Pressing and holding

At the pressing-and-holding step, the resin is heated over the resin-glass transition temperature (T_g) and the resin is assumed to be a rubber elastic body. Figure 6.11a shows the stress distribution in the resin during the step. The compulsive stress is induced underneath the convex portion of the mold. On the other hand, the resin along the mold sidewall boundary is stretched, but the stress is not too large. Anyway, we cannot find a specific origin of defects from this stress distribution because the resin is soft and the stress is spread over the resin.

6.4.2.2 Step 2: Cooling

The stress at the end of the cooling step is shown in Fig. 6.11b. The resin is cooled to below T_g, and the resin turns into an elastic body, where the pressure is still applied until cooling is completed. In this case, the resin is assumed to be an elastic material. The simulated result shows that a large stress concentration is induced at the corner of the pattern, which may induce a fatal crack in the resin.

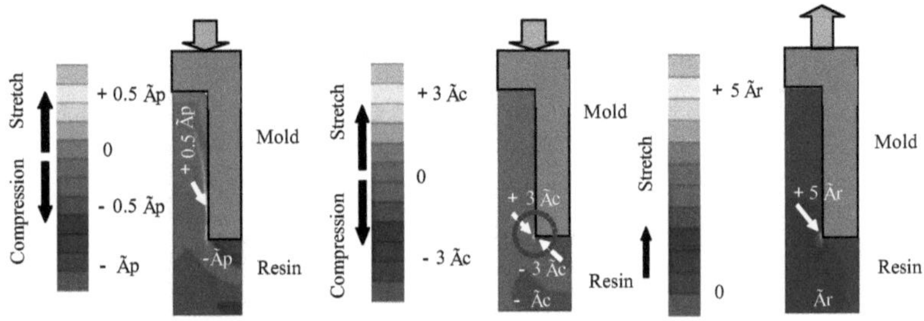

Figure 6.11 Principal stress distribution.

6.4.2.3 Step 3: Releasing

Figure 6.11c shows the stress distribution in the release process. In the releasing step, the resin is stretched by a static friction force between the mold and the resin. The stress is again concentrated at the corner of the pattern.

As discussed above, fatal incidents of the fracture defects are theoretically investigated by simulation studies. The fracture defect is induced during the cooling processes in the current process sequence.

6.4.3 Optimization of the Process Sequence

To eliminate the defects, the revised process sequence is examined. The imprinting pressure is gradually released during the cooling step, as shown in Fig. 6.12a, because the applied pressure below the glass transition temperature of the resin induces stress concentration. In the revised sequence, the cooling time is also extended to release the residual stress. Figure 6.12b shows the experimental results using the revised sequence. The line patterns are successfully fabricated.

These results support the simulation studies as discussed above.

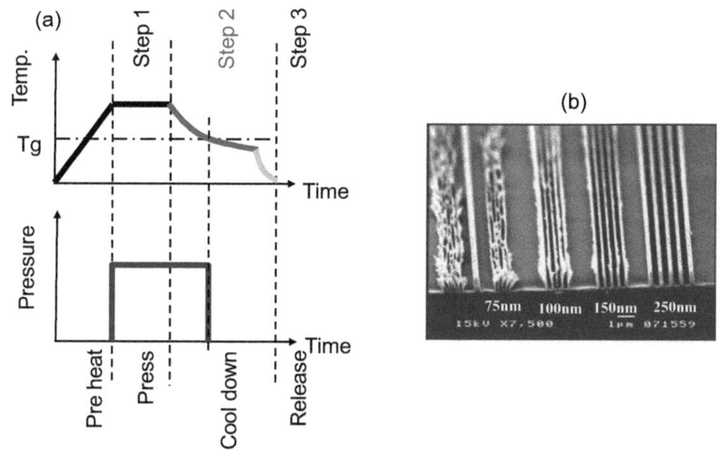

Figure 6.12 Process experimental result for high-aspect-ratio pattern fabrication by the advanced imprint process, where the applied pressure is released below T_g and a slow cooling sequence. (a) Revised process sequence. (b) SEM image of the experimental result. SEM, scanning electron microscopy.

6.4.4 High-Aspect-Ratio Pattern Fabrication [5]

According to the above discussions, a high-aspect-ratio pattern is fabricated by the thermal nanoimprinting process. Figure 6.13 demonstrates high-aspect-ratio pattern fabrication, where the aspect ratio is 8.6 for 100 nm lines. The groove depth of the mold is almost the same as the imprinted pattern height, and the initial resin thickness is over four times the mold groove depth. In this case, nanoimprinting is done under atmospheric air ambient.

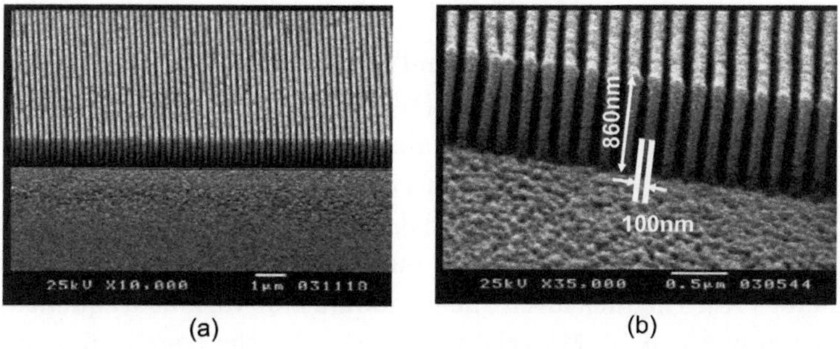

Figure 6.13 High-aspect-ratio pattern fabrication by nanoimprint lithography using thick resin. (4 μm PMMA–based resin on Si, 170°C, 20 MPa, Mw = 100 k). (a) Low-magnification SEM image. (b) High-magnification SEM image.

6.5 Time-Dependent Analysis [7, 8]

In this section, the time-dependent resin deformation process is discussed. It is essential to know the time evolution of resin deformation and what happens in the process in order to obtain the optimum process conditions for cost-effective NIL. However, the resin deformation processes have not been clearly revealed because of the difficulties of evaluation in the resin time evolution process. To obtain the time evolution of resin profiles experimentally, we performed a series of novel experiments for various process (hot pressing) times by careful observation of the quenched resin profiles.

On the other hand, we have recently developed software for the numerical simulation to investigate the time-dependent resin deformation process and the thermal cooling process in T-NIL and evaluate the validity of the simulation model.

Experiments and simulation studies are carried out to investigate the time evolution of resin deformation and dependency of pattern size on the pressing process time.

6.5.1 Numerical Models

To simulate resin deformation in the T-NIL processes, we recently developed a finite element program. The program was designed to calculate the resin deformation in typical T-NIL processes. The constitutive model of the resin is the viscoelastic model based on the generalized Maxwell [14–16] model, as shown in Fig. 6.14a.

The complex instantaneous shear modulus G_0 is expressed as

$$G_0 = G_{\mathrm{inf}} + G_1 + G_2 + \ldots, \tag{6.9}$$

where G_i is the shear modulus and G_{inf} is a long-term shear modulus.

The Williams–Landel–Ferry (WLF) law [17] is applied to consider its temperature dependency, as shown in Figs. 6.14b and 6.14c.

The Si mold was modeled as a rigid body and the PMMA resin was modeled as a viscoelastic material, as described above. The material properties of the PMMA were measured using a rotary rheology meter [18]. The measured shear modulus for various frequencies is shown in Fig. 6.14. Material parameters for the constitutive model were obtained from the measurement data through certain data conversion.

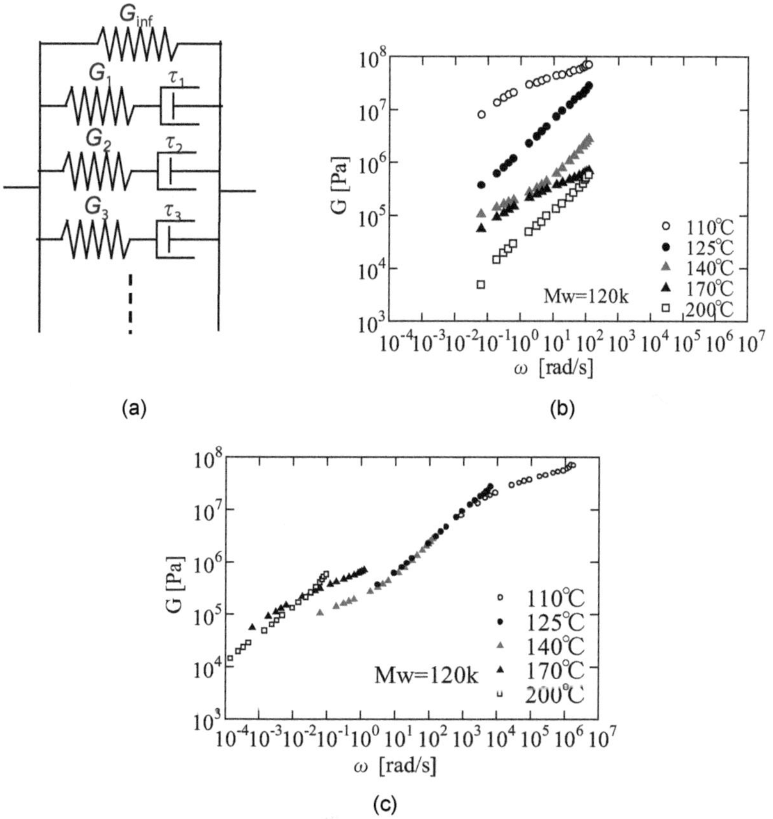

Figure 6.14 Viscoelastic model for time dependent analysis. (a) The schematic diagram of the generalized Maxwell model. (b) Examples of the measurement results for shear models G. (c) Examples of the master curve obtained by WLF expansion. τ_i is the relaxation time of the *i*th component.

6.5.2 Experimental Study

The schematic diagram of the experiment is shown in Fig. 6.15. The mold has L&S patterns 1200 nm in line width, 300 nm in space width, and 240 nm in height. PMMA is used, and the imprint temperature was 140°C. The imprint pressure was 5 MPa. The process time was varied alternatively from 10 s to 60 s. After the resin was pressed for a specific process period (process time), the sample with the mold and the substrate were quickly removed from the thermal stage and put on a thermally controlled Al cooling plate. As the Si substrate and the Si mold are 0.6 mm in thickness and have high thermal conductivity, the sample must be cooled down below the resin–glass transition temperature within 1 s. Finally, the mold was removed from the resin. As a result, the resin retained its exact shape after thermal press.

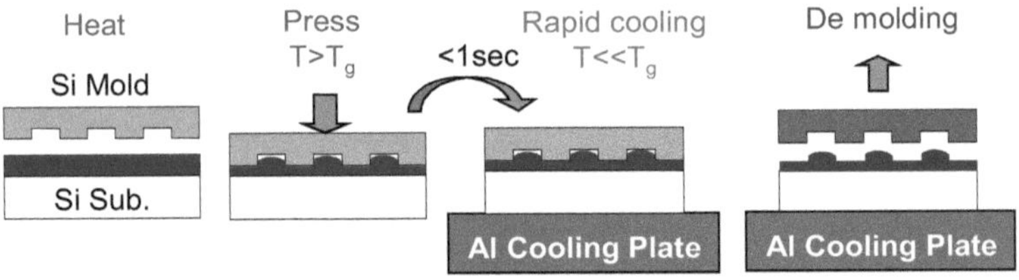

Figure 6.15 Schematic diagram of the experiment procedure.

6.5.3 Result and Discussion

The time evolutions of the resist cross-sectional profiles in both simulation and experimental results are shown in Fig. 6.16 for various values of pressure.

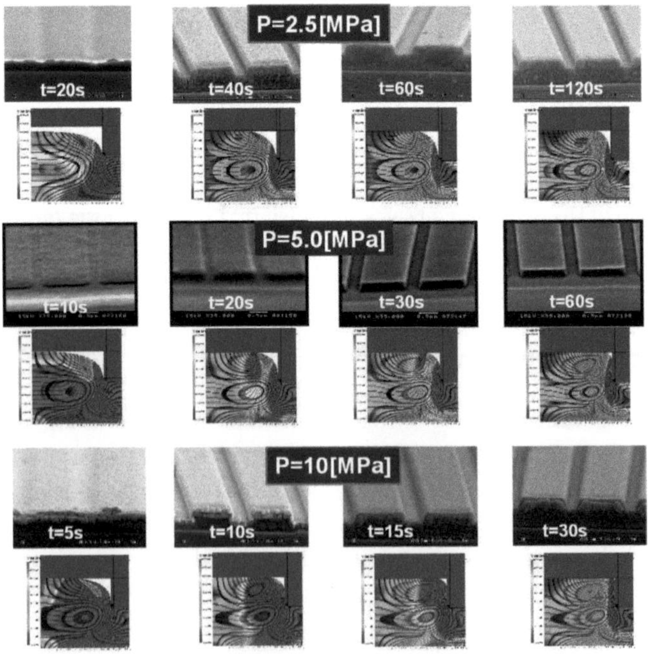

Figure 6.16 Simulation and experimental results for the time evolution profiles of the resist for various imprinting pressures.

When the imprint pressure increases, the corresponding process times are inversely proportional in time evolution profiles.

This is understood by the following dimensional study. The viscosity η [Pa s] is expressed by multiplying pressure and time. So, the process time τ [s] is expressed by

$$\tau \propto \eta/P, \tag{6.10}$$

where P [Pa] is the imprint pressure.

The simulation and experimental results well support the above approximation, which is an important relation between process time and process condition as applied pressure.

6.6 Summary

The resin deformation process is investigated by numerical simulations and compared to the experimental results.

The required pressure depends on the aspect ratio of the pattern. When the aspect ratio increases, the required pressure increases because the resin is hardly filling the narrow cavity. On the other hand, when the aspect ratio decreases to a value below 0.8, the required pressure again increases. This is because the resin is deformed only near the cavity edge area and not easily deformed at the center area of the cavity. As a result, large pressure is required.

When the initial thickness becomes less, the required pressure increases because the resin is fixed on the substrate and deforming becomes difficult near the substrate. On the other hand, when the initial thickness increases, the resin is easily deformed and the center portion in the cavity pattern is bunched up from the surrounding resin.

About the process time, the deformation time is almost inversely proportional to the applied pressure.

These computational results fairly agree with the experimental results. The results give useful suggestions for process design and optimization.

Finally, what is important to notice is that all computational results depend on geometrical configuration but not absolute feature size under the continuum approximation, which may be effective down to several tenths of nanometers, as large as the polymer size of the resin. Near polymer size, we need molecular dynamics studies in nanoscale area.

References

1. S. Y. Chou, P. R. Krauss, and P. J. Renstrom (1995). Imprint of sub-25 nm vias and trenches in polymers, *Appl. Phys. Lett.*, **67**, 3114.
2. H. Scheer, H. Schultz, T.Holffmann, and C.Torres (1998). Problems of the nanoimprinting technique for nanometer scale pattern definition, *J. Vac. Sci. Technol.*, **B16**, 3917.
3. L.J. Heyderman, H. Schift, C. David, J. Gobrecht, and T. Schweizer (2000). Flow behaviour of thin polymer films used for hot embossing lithography, *Microelectron. Eng.*, **54**, 229.
4. X. Chung and L.J. Guo (2000). One step lithography for various size patterns with a hybrid mask-mold, *Microelectron. Eng.*, **71**, 288.
5. Y. Hirai, T. Konishi, T. Yoshikawa, and S. Yoshida (2004). Simulation and experimental study of polymer deformation in nanoimprint lithography, *J. Vac. Sci. Technol.*, **B22**, 3288.
6. Y. Hirai, S. Yoshida, and N. Takagi (2003). Defect analysis in thermal nanoimprint lithography, *J. Vac. Sci. Technol.*, **B21**, 2765.
7. Y. Hirai, Y. Onishi, T. Tanabe, M. Nishihata, T. Iwasaki, H. Kawata, and Y. Iriye (2007). Time dependent analysis of the resist deformation in thermal nanoimprint, *J. Vac. Sci. Technol. B*, **25**, 2341.
8. Y. Hirai, Y. Onishi, T. Tanabe, M. Shibata, T. Iwasaki, and Y. Iriye (2008). Pressure and resist thickness dependency of resist time evolutions profiles in nanoimprint lithography, *Microelectron. Eng.*, **85**, 842.
9. Y. Hirai, M. Fujiwara, T. Okuno, Y. Tanaka, M. Endo, S. Irie, K. Nakagawa, and M. Sasago (2001). Study of the resist deformation in nanoimprint lithography, *J. Vac. Sci. Technol.*, **B19**, 2811.
10. For example, L. H. Sperling (2001). *Introduction to Physical Resin Science*, Wiley-Interscience, NJ, USA.
11. M. Mooney (1940). A theory of large elastic deformation, *J. Appl. Phys.*, **11**, 582.
12. R. S. Rivlin (1949). Large elastic deformations of isotropic materials, *Phil. Trans. R. Soc.*, **A242**, 173.
13. User Information A, MARC Analysis Research Corporation (1994) Palo Alto.
14. W. B. Young (2005). Analysis of the nanoimprint lithography with a viscous model, *Microelectron. Eng.*, **77**, 405.

15. T. Eriksson and H. K. Rasmussen (2005). The effects of polymer melt rheology on the replication of surface microstructures in isothermal moulding, *J. Non-Newtonian Fluid Mech.*, **127**, 191.
16. T. A. Osswald and H. L. Menges (1996). *Material Science of Polymers for Engineers*, Hanser, New York.
17. M. Williams, R. Landel, and D. Ferry (1955). The temperature dependence of relaxation mechanisms in amorphous polymers and other glass-forming liquids, *J. Am. Chem. Soc.*, **77**, 3701.
18. Y. Hirai, T. Yoshikawa, N. Takagi, and S. Yoshida (2003). Mechanical properties of poly-methyl methacrylate (PMMA) for nano imprint lithography, *J. Photopolym. Sci. Technol.*, **16**, 615.

Chapter 7

UV Nanoimprint Lithography Process Simulation

Yoshihiko Hirai
Department of Physics and Electronics Engineering,
Graduate School of Engineering, Osaka Prefecture University,
Gakuen-cho, Sakai, 599-8531 Osaka, Japan
hirai@pe.osakafu-u.ac.jp

7.1 Introduction

Ultraviolet nanoimprint lithography (UV-NIL) [1] has an advantage in process compatibility with conventional semiconductor lithography. In the UV-NIL, a UV-curable resin is used and cured by UV exposure. Unlike thermal NIL (T-NIL), the resin used in UV-NIL is limited but the process time is short and the process could be done at room temperature. The resin is sometimes called UV resist.

UV-NIL needs several process steps. Figure 7.1 shows the process steps and the essential mechanisms in each step. The first step is the process of the resin filling into the pattern cavity, where the resin used in UV-NIL has a lower viscosity than that used in T-NIL.

Nanoimprint and Its Applications
Edited by Akihiro Miyauchi

ISBN 978-981-4800-37-2 (Hardcover), 978-0-429-03192-2 (eBook)
www.jennystanford.com

So, the filling mechanism of the resin into the mold cavity in UV-NIL is different than that in T-NIL. Next, UV is irradiated to cure the UV resin. When the optical index of the resin is slightly different from that of the mold, we should take care of the UV wave propagation into the resin. Also, the curing characteristics of the resin should be considered to optimize the process conditions. After curing is completed, the mold is removed.

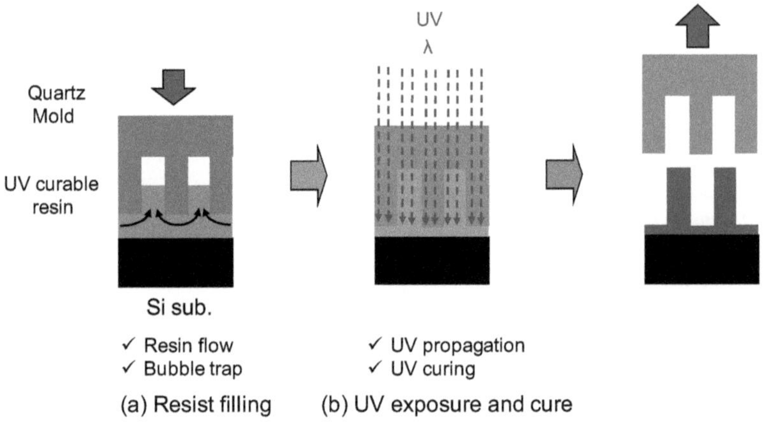

Figure 7.1 Schematic diagram of the UV-NIL process and simulation systems.

Understanding these process physics and chemistry is essential to optimize the UV-NIL process and eliminate defects.

In this chapter, basic sciences in UV-NIL are briefly introduced and the process simulation for UV-NIL based on conventional continuous mechanics is demonstrated to discuss the process mechanism [2].

7.2 Resist-Filling Process [3, 4]

7.2.1 Numerical Model

In the UV nanoimprinting process, two ways have been proposed for the resist-coating process: a resist droplet process and a spin-coating process.

In both processes, air bubbles could be trapped in the pattern cavities of molds [5]. Simulating the resist flows helps understand the resist-filling mechanism. The resist is assumed to be an incompressible fluid, and the Navier–Stokes equations and continuous equation are applied.

$$\rho\frac{\partial\vec{v}}{\partial t}+\rho\left(\vec{v}\cdot\text{grad}\right)\vec{v}=-\text{grad}\,P+\rho\vec{g}+\eta\left(\text{grad}\,\text{div}\,\vec{v}\right) \tag{7.1}$$

$$\text{div}\,\vec{v}=0 \tag{7.2}$$

Also, the problem of air compression in a cavity is solved, simultaneously.

Here, ρ, η, and g are the density, the viscosity, and the gravitational acceleration, respectively. In this work, commercially available software U-FLOW@/AG (Mizuho Information & Research Institute) is used, based on the adaptive mesh refinement technique [6], to solve the resist and air flow.

For boundary conditions, the contact angle should be determined from the balance of the surface energy at the contact point and depends on the flow condition of the liquid. However, the determination of contact angle involves a lot of computational costs. Thus, mostly fluid software gives the contact angle as a process parameter under quiescent conditions instead of giving the surface energy.

7.2.2 Resist Droplet Process under Air Ambient

Figure 7.2 shows a schematic of a resist droplet process in UV-NIL. A resist is dropped on a substrate, and a mold with fine pattern cavities is pressed on it. The resist is spread laterally into the gap between mold and substrate and filled into the cavity pattern. The mold and the substrate are set in a fixed position, and the resist is injected into the space by a constant velocity $\vec{v}$.

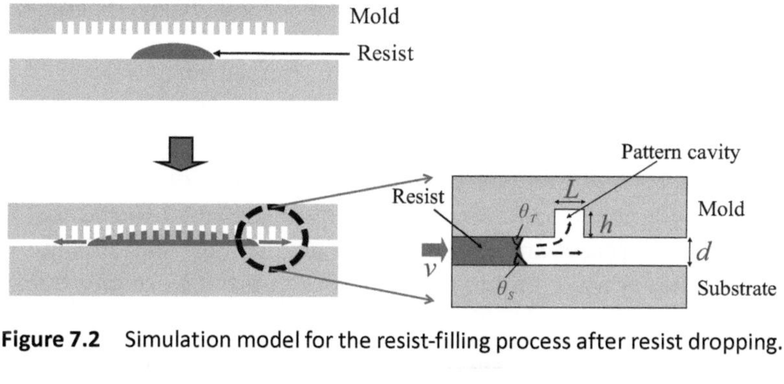

Figure 7.2 Simulation model for the resist-filling process after resist dropping.

To express the surface tension, the static contact angles between the resist and the mold θ_T and the resist and the substrate θ_S are given.

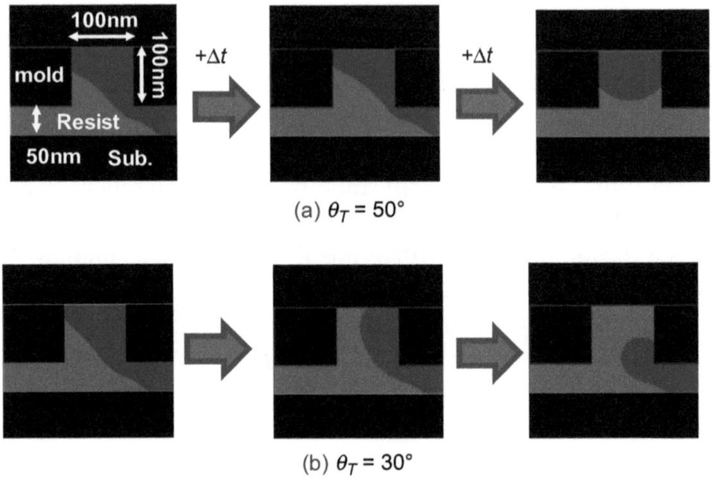

Figure 7.3 Simulation result of resist flow at various contact angles for the mold θ_T (L = 100 nm, h = 100 nm, d = 50 nm, $\vec{v}$ = 0.01 m/s, η = 10 mPa·s).

Figure 7.3 shows simulation results. The resist flows at various contact angles for the mold θ_T. When the θ_T is large and the θ_S is small, the resist flow along the mold is suppressed and the resist flows laterally. As a result, the resist does not flow into the cavity, as shown in Fig. 7.3a, and air is trapped.

On the other hand, when the θ_T is small, the resist easily flows along the mold surface and the resist pushes the air out of the cavity. As a result, the resist is filled into the cavity without air getting trapped, as shown in Fig. 7.3b. The situation depends on cavity size, contact angles, viscosity, etc.

Figure 7.4 shows the map for air trapped at various contact angles to the mold θ_T and the substrate θ_S when the viscosity η of the resist is 10.0 mPa·s.

As the contact angle to the mold θ_T decreases, the bubble defects can be eliminated. However, the adhesion between the resist and the mold increases and the demolding process becomes difficult. So, the mold surface condition should be optimized to reduce bubble trapping.

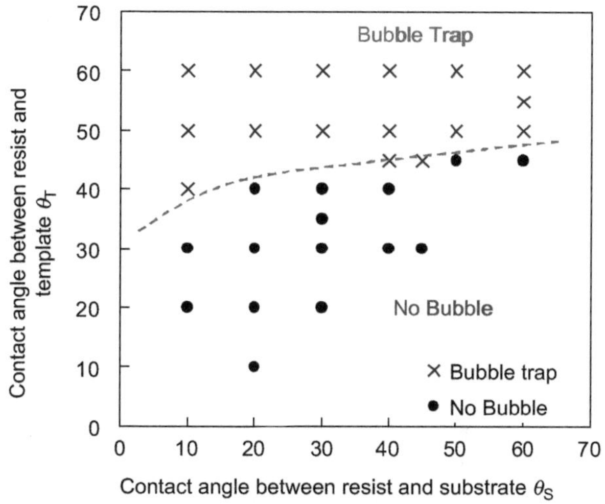

Figure 7.4 The map for air trapped at various contact angles to the mold θ_T and the substrate θ_S (L = 100 nm, h = 100 nm, d = 50 nm, $\vec{v}$ = 0.01 m/s, η = 10 mPa·s).

7.2.3 Resist Spin-Coating Process under Condensable Gas Ambient

A resist spin-coating process is profitable for large-field processing because the resist quickly spreads over a large field. However, air is trapped in the cavity [11, 15], as shown in Fig. 7.5.

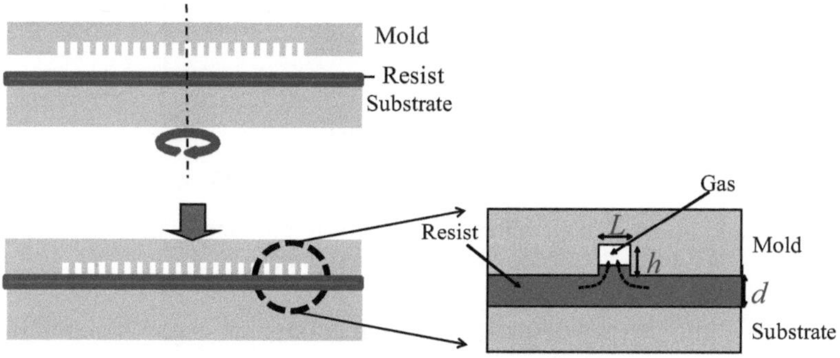

Figure 7.5 Schematic of the resist spin-coating process.

The air sometimes dissolves in the resist when the trapped air is highly compressed, but usually air bubbles remain.

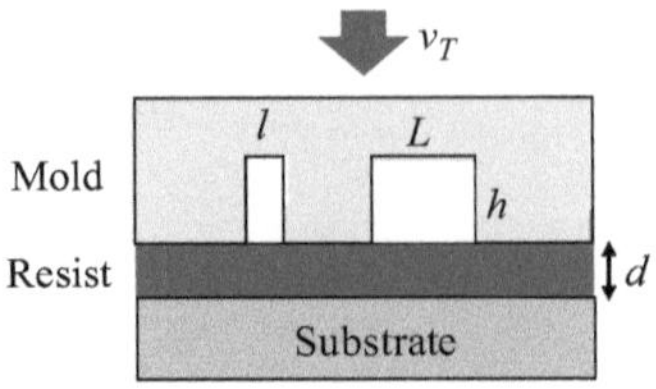

Figure 7.6 Simulation model of the resist-filling process for spin-coated resist.

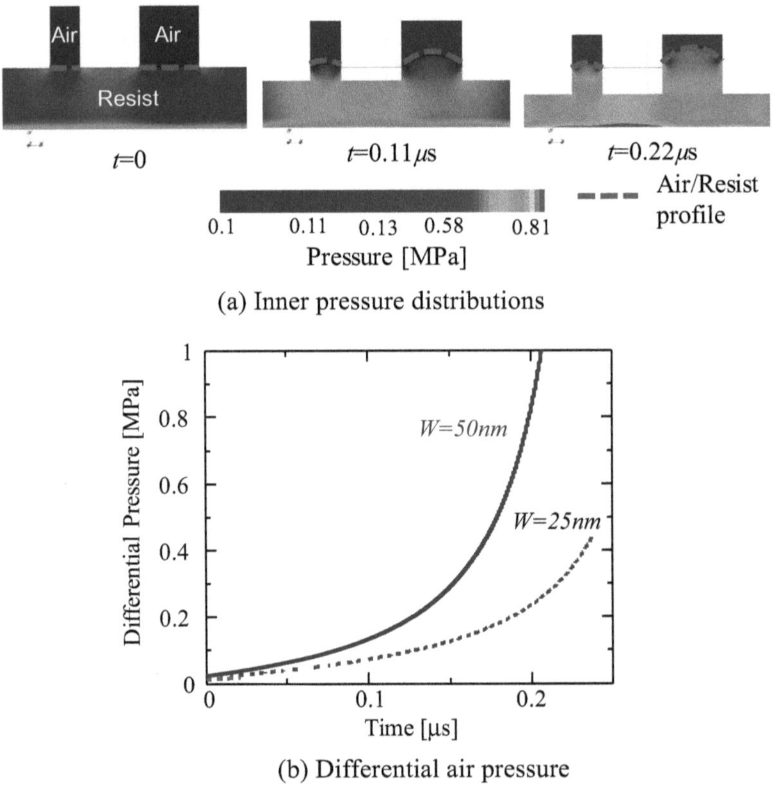

Figure 7.7 Resist-filling profiles and pressure in resist and air (η = 50 mPa·s, θ_T = 60°, $\vec{v}_T$ = 0.1 m/s, l = 25 nm, L = 50 nm, h = 50 nm).

To eliminate the defects, the use of condensable gas is one of the promising solutions.

The schematic of the simulation model is shown in Fig. 7.6. The mold presses the resist by a constant velocity $\vec{v}_T$ over the spin-coated resist. The narrow line (l = 25 nm) and the wide line

(L = 50 nm) are periodically placed. The resist fills into the narrow pattern more slowly than into the wide pattern due to surface tension of the resist, as shown in Fig. 7.7a. The differential pressure in each cavity is also shown in Fig. 7.7b. The air in the wide pattern is compressed faster than that in the narrow pattern.

To eliminate trapping of air, one promising way is to use a condensable gas, such as pentafluoropropane (PFP) [7], which turns into a liquid under compression (over around 0.15 MPa) and so the bubble disappears.

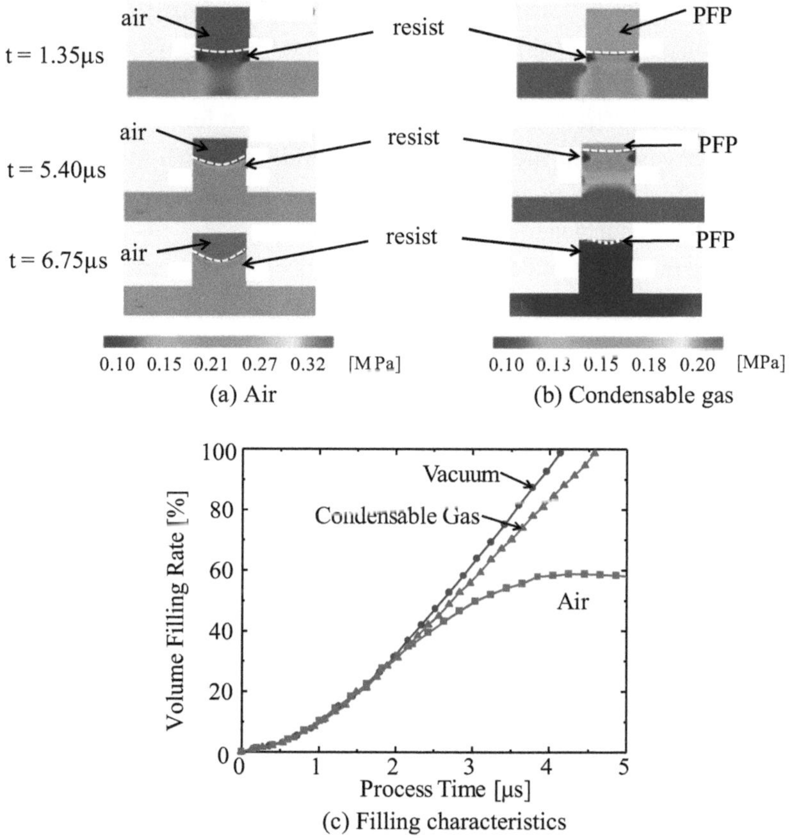

Figure 7.8 Time evolutions of the resist-filling process into cavity under air and condensable gas atmospheres. (a) Air, (b) condensable gas (PFP), and (c) resist-filling characteristics (η = 50 mPa·s, θ_T = 60°, $\vec{v}_T$ = 0.1 m/s, L = 50 nm, h = 140 nm).

Figures 7.8a and 7.8b show the simulation results of the resist-filling process in air and in condensable gas ambient. The colored maps show inner pressure distributions. In the case of air ambient, the trapped air is compressed. On the other hand, the trapped gas pressure remains constant and the gas finally disappears as the mold is pressed. In the air ambient, the mold is stopped midway where the air pressure and applied force are balanced.

Figure 7.8c shows the resist-filling characteristics in various ambients. In the condensable gas ambient, the bubble is eliminated and the resist is fully filled into the cavity. The resist-filling time in condensable gas is almost the same as that in vacuum ambient.

The impact of the resist-filling time on material characteristics is investigated. The resist-filling time is simulated for various contact angles θ_T between the resist and the mold, as shown in Fig. 7.9. As the contact angle θ_T increases, the filling time increases in narrow lines due to surface tension at the boundary.

The method is effective for the resist droplet process to eliminate the bubbles.

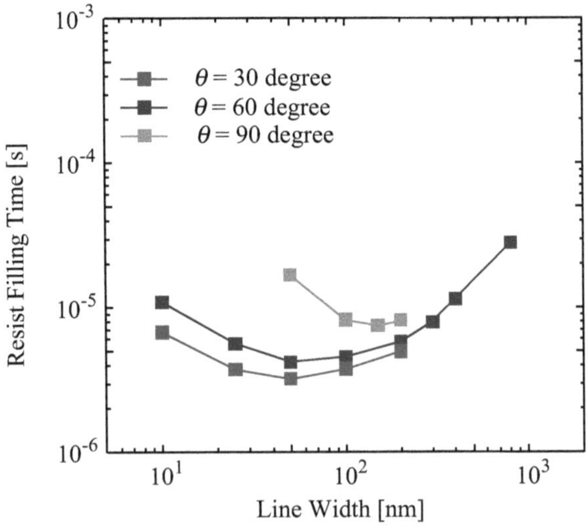

Figure 7.9 Resist-filling time with variation in the contact angle between resist and mold (h = 100 nm, η = 50 mPa·s).

7.3 UV Exposure Process [8]

7.3.1 Numerical Model

To calculate optical intensity distribution in the UV-curable resist, electromagnetic propagation of the irradiated light in various configurations was analyzed by numerically solving Maxwell's equation, as shown in Eqs. 7.3 and 7.4, using the finite-difference time-domain method [9].

$$\varepsilon\frac{\partial E(r,t)}{\partial t}=\nabla\times H(r,t) \tag{7.3}$$

$$\mu\frac{\partial H(r,t)}{\partial t}=-\nabla\times E(r,t) \tag{7.4}$$

Figure 7.10 shows a schematic diagram of the UV exposure process. The photopolymer is filled between a quartz mold and a Si substrate. The light beam is irradiated on the photopolymer by a single-wavelength plane wave optical source. In this system, the optical transmittance of the photopolymer is assumed to be 100%. The wavelength λ of the irradiated beam and the optical index of the quartz mold are assumed to be 320 nm and 1.49, respectively.

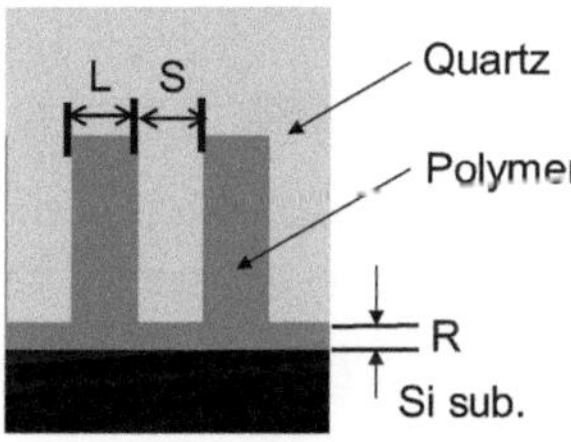

Figure 7.10 Schematic diagram of the analysis system for UV exposure.

7.3.2 Simulation Models and Demonstrations

7.3.2.1 Impact of feature size

The line width L is varied from 320 nm (L/λ = 1) to 80 nm (L/λ = 1/4). The optical index of the polymer is assumed to be 1.6. The residual thickness of the polymer is assumed to be 100 nm, and the groove depth of the mold (pattern height) is set to be 480 nm.

Figure 7.11 shows optical intensity distributions in the polymer for various line widths. When the line width is equal to the wave length (L = 320 nm), the irradiated beam propagates mostly in the polymer region, where the optical index is higher than that of the quartz mold. Also, a standing wave is induced. When the line width becomes narrow, the light beam seeps out of the resist. This is because the propagation of the light wave is not influenced by the optical index difference in ultrafine structures, which is sufficiently smaller than the wavelength of light.

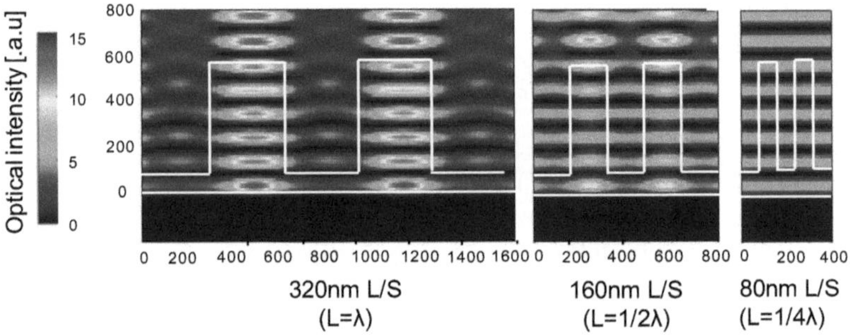

Figure 7.11 Optical intensity distribution for various pattern feature sizes (λ = 320 nm, optical index of the quartz n_q = 1.49, optical index of the photopolymer n_p = 1.6).

7.3.2.2 Impact of optical index

To investigate influence of optical characteristics of UV resist, the optical index is varied from 1 to 1.8, where the residual thickness of the resist is assumed to be 0 and the optical index of the quartz mold is 1.49, as shown in Fig. 7.12.

When the index of UV resist is smaller than that of the quartz mold, the irradiated light beam passes through the quartz mold and does not expose the UV resist. This means that the optical index of the resist should be equal to that of the mold but we must use a higher-optical-index resist than that of mold materials.

Also, optical diffraction and interference sometimes cause fatal defects. On the other hand, the optical intensity affects the resist curing process, as discussed below.

Optical intensity in the UV resist is important to predict fatal defects, especially using coherency optical sources such as a light-emitting diode (LED).

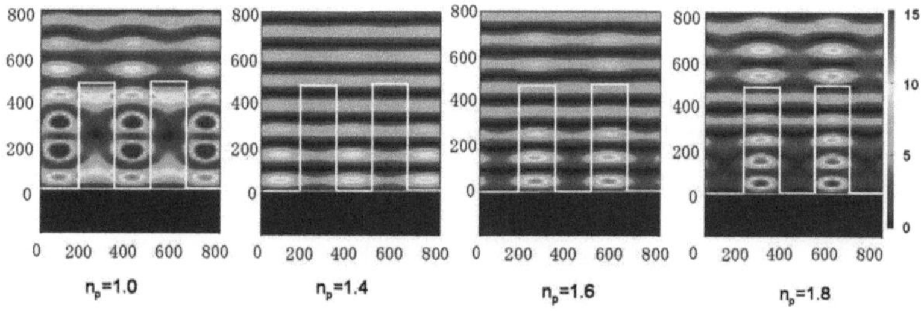

Figure 7.12 Optical intensity distribution for various indexes of the resist.

Careful investigation is demanded of mold design and process conditions to avoid defects.

7.4 UV Curing Process [10, 11]

7.4.1 Numerical Model

In the UV nanoimprinting process, the UV curing process condition is important for process design such as the UV exposure time. In this chapter, basic kinetics of UV polymerization [12] are introduced using an experimental study.

Firstly, the photoinitiator C_i generates a radical R^* by UV irradiation.

$$C_\mathrm{i} \rightarrow R^* \tag{7.5}$$

Then, the radical reacts with a monomer M in a resin and turns into a radical monomer $R - M^*$.

$$R + M \rightarrow R - M^* \tag{7.6}$$

The activated radical monomer reacts with another monomer in a chain reaction.

$$R - M_n{}^* + M \rightarrow R - M_{n+1}{}^* \tag{7.7}$$

Finally, they react with each other and the chain reactions are terminated.

$$R - M_m{}^* + R - M_n{}^* \rightarrow R - M_{m+n} - R \tag{7.8}$$

Here, the initiation reaction velocity $\vec{V}_\mathrm{R}$ is expressed as

$$\vec{V}_\mathrm{R} = \frac{d[R]}{dt} = 2k_\mathrm{d} f[C_\mathrm{i}], \tag{7.9}$$

where K_d, f, $[R]$, and $[C_i]$ are the reaction velocity constant, the coefficient of UV absorption, the radical concentration, and the initiator concentration, respectively.

On the other hand, the polymerization reaction velocity $\vec{V}_P$ is expressed as

$$\vec{V}_p = -\frac{d[M]}{dt} = k_p[R][M], \tag{7.10}$$

where k_p and $[M]$ are the reaction rate constant and the monomer concentration, respectively. Also, the termination velocity $\vec{V}_t$ is expressed as

$$\vec{V}_t = k_t[R][R], \tag{7.11}$$

where k_t is the reaction rate constant.

In steady state, $\vec{V}_R$ and $\vec{V}_t$ are balanced with each other. Thus, the radical concentration $[R]$ is expressed as

$$[R] = \left[\frac{2k_d f}{k_t}\right]^{1/2} \sqrt{[C_i]} \tag{7.12}$$

Substituting Eq. 7.8 for Eq. 7.6, V_p is expressed as

$$\begin{aligned} V_p &= -\frac{d[M]}{dt} = k_p[R][M] \\ &= k_p\left[\frac{2k_d f}{k_t}\right]^{1/2} [M]\sqrt{[C_i]} \\ &= K[M]\sqrt{[C_i]} \end{aligned} \tag{7.13}$$

Here $K = k_p\left[\frac{2k_d f}{k_t}\right]^{1/2}$.

As a result, time-dependent monomer concentration is expressed as

$$-\frac{d[M]}{dt} = K[M]\sqrt{[C_i]} \tag{7.14}$$

When enough photoinitiator is contained in the resin and the degradation is ignored ($[C_i]$ = constant), the monomer concentration becomes

$$[M] = [M_0]\exp(-K\sqrt{[C_i]}\cdot t), \tag{7.15}$$

where $[M_0]$ is the initial concentration of the monomer. As a result, the monomer conversion ratio CR is expressed as

$$\mathrm{CR} = 1 - \frac{[M]}{[M_0]} = 1 - \exp(-K\sqrt{[C_i]}\cdot t) \tag{7.16}$$

On the other hand, the rate of photochemical initiation is given by

$$R_i = 2\Phi I_a, \tag{7.17}$$

where I_a is the intensity of the absorbed light in moles of light quanta and Φ is the number of propagating chains initiated per light photon absorbed. Then, the photoinitiator concentration $[C_i]$ is given as

$$[C_i] \propto 2\Phi I \propto [C_o]I, \tag{7.18}$$

where $[C_o]$ is the concentration of the primary initiator in the resin. As a result, the conversion ratio becomes

$$\mathrm{CR} = 1 - \exp(-\sqrt{I}\cdot t/\tau). \tag{7.19}$$

The conversion is related to $\sqrt{I}\cdot t$ and τ. We define $\sqrt{I}\cdot t$ and τ as the effective conversion time and the related time constant of the polymerization, respectively.

7.5 Experimental Results

7.5.1 Photoinitiator Concentration during UV Exposure

In the above discussions, in Eq. 7.11, enough photoinitiator is supplied by the primary initiator and $[C_i]$ is considered to be constant. To verify the assumption, the concentration of the photoinitiator is observed by UV absorption spectrum USB4000-UV-VIS (Ocean Optics).

Figure 7.13 shows the UV absorbance spectrum for C-TGC-02 resist (TOYO GOSEI), which is a test rig rein containing an acrylic acid adduct of propylene glycol diglycidyl ether (epoxy ester 70PA) as the polymerization resin with 2-methyl-1-(4-methylthiophenyl)-2-morpholinopropane-1-one (IRGACURE907) as the UV polymerization initiator.

The UV absorbance is almost constant below around 100 mJ/cm^2 in dosage. This means the primary initiator supplies sufficient photoinitiator, the concentration of the photoinitiator is almost constant ($[C_i]$ = constant), and Eqs. 7.15 and 7.19 are approved below 100 mJ in UV dosage.

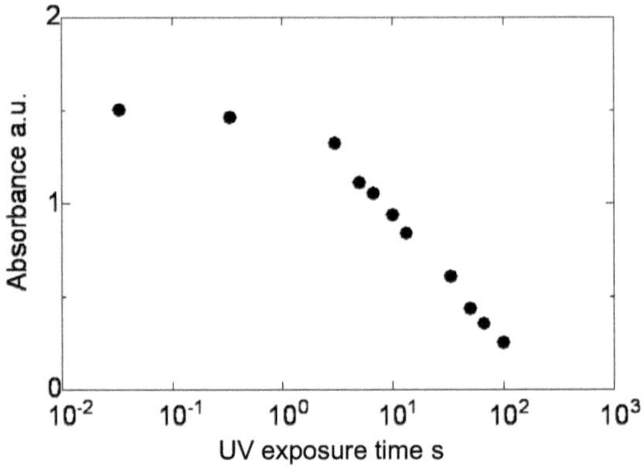

Figure 7.13 UV absorbance of the primary initiator of the C-TGC-02 resist (I = 30 mW/cm^2, λ = 320 nm).

7.5.2 Monomer Conversion

To evaluate the monomer conversion in the C-TGC-02 resist, cleavage between the double bond of carbons in the monomer is observed using Fourier transform infrared spectroscopy. The cleavage occurs at 810 cm^{-1} in wave number. Figure 7.14 shows the measured conversion ratio of the monomer for various UV intensities using PAGA-100 (Lithotech Japan) under inert gas ambient to avoid oxygen inhibition versus effective conversion time $\sqrt{I}\cdot t$. In those cases, the consumption energy (2 mJ) for a photostabilizer to inhibit the chain reaction is subscripted. The monomer conversion ratios for various UV intensities are expressed as a function of $\sqrt{I}\cdot t$, as explained in Eq. 7.15. This is a very important characteristic and a big difference from the conventional photoresist. The curing characteristic is not determined by the UV exposure energy $I{\cdot}t$ (dosage) but $\sqrt{I}\cdot t$. So, if you want to reduce the exposure time to 1/2, you should increase the UV intensity to 4 times.

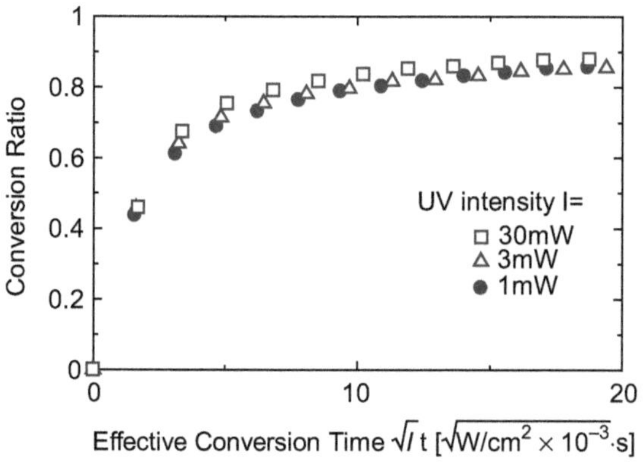

Figure 7.14 Relation between effective conversion time $\sqrt{I}\cdot t$ and monomer conversion ratio (C-TGC-02:10 μm).

7.5.3 Elastic Modulus

The modulus of the UC resist during UV exposure is measured using a rheometer. We use UV rheometer MCR301 (Anton Paar). Figure 7.15 shows the result. The characteristics of an elastic storage modulus also depend on the effective conversion time $\sqrt{I}\cdot t$. Figure 7.16 shows the relationship between the conversion ratio and the storage modulus for equivalence effective conversion time $\sqrt{I}\cdot t$. As shown in Fig. 7.16, there exists a correlation among them, as expressed in

$$G' = G_0 + G_b \cdot CR^{\alpha} \tag{7.20}$$

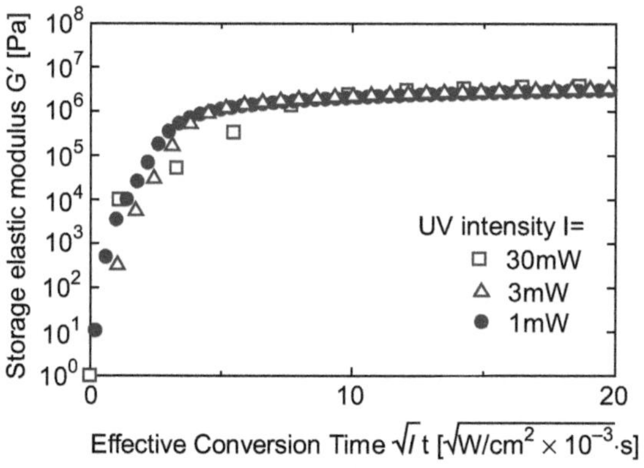

Figure 7.15 Relation between effective conversion time $\sqrt{I}\cdot t$ and storage elastic modulus (C-TGC-02:10 μm).

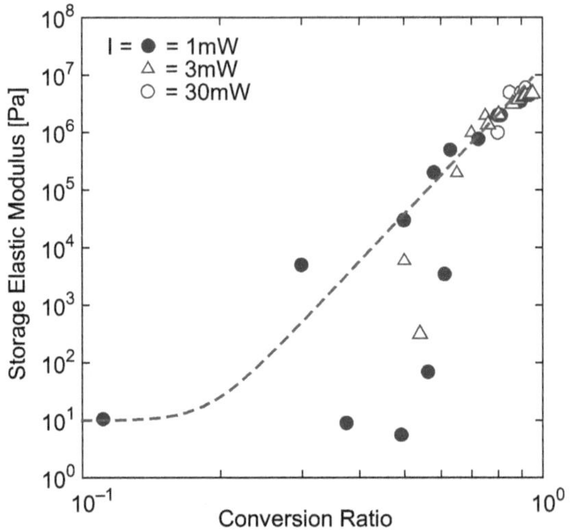

Figure 7.16 Relation between conversion ratio and elastic modulus.

The extracted values of G_0, G_b, and α are 10 Pa, 1.4×10^7, and 8.5, respectively. This shows that the resist modulus depends on the conversion ratio.

7.6 Summary

Modeling and numerical simulations for the UV nanoimprinting process are demonstrated based on conventional contentious mechanics and polymer chemistry. Understanding those phenomena helps in the investigation of process defects and optimization of process conditions and material properties.

In the resist-coating process, the combination of the surface tensions between mold and substrate is a key issue to avoid bubble trapping. However, bubble trapping is an unavoidable event in UV nanoimprinting. To eliminate bubble trapping, use of a condensable gas was confirmed by simulation work.

In the UV exposure process, we should take care of optical propagation, such as interference and diffraction, in mold and resist. Especially when you use LEDs as the UV source, the LEDs are highly coherent and could cause interference and diffraction.

In the UV curing process, we should take care that the polymerization by UV cure is determined by the product of exposure time and the square root of UV intensity, not dosage as the product of exposure time and UV intensity.

Acknowledgments

The author conveys heartfelt thanks to Dr. T. Iwasaki at the Mizuho Information and Research Institute and Dr. S. Ishikawa at Mechanical Design Co. Ltd. for their helpful support in building the systems. Also, the author conveys heartfelt thanks to Dr. H. Hiroshima at the National Institute of Advanced Industrial Science and Technology and Dr. N. Sakai at Toyo Gosei Co. Ltd. for their helpful discussions on numerical modeling.

References

1. M. Colburn, S. Johnson, M. Stewart, S. Damle, T. Bailey, B. Choi, M. Wedlake, T. Michaelson, S. V. Sreenivasan, J. Ekerdt, and C.G. Willson (1999). Step and flash imprint lithography: a new approach to high-resolution patterning, *Proc. SPIE*, **3676**, 378.
2. M. Shibata, A. Horiba, Y. Nagaoka, H. Kawata, M. Yasuda, and Y. Hirai (2010). Process-simulation system for UV-nanoimprint lithography, *J. Vac. Sci. Technol.* B, **28**, C6M108.
3. D. Morihara, Y. Nagaoka, H. Hiroshima, and Y. Hirai (2009). Numerical study on bubble trapping in UV nanoimprint lithography, *J. Vac. Sci. Technol. B*, **27**, 2866.
4. Y. Nagaoka, R. Suzuki, H. Hiroshima, N. Nishikura, H. Kawata, N. Yamazaki, T. Iwasaki, and Y. Hirai (2012). Simulation of resist filling properties under condensable gas ambient in ultraviolet nanoimprint lithography, *Jpn. J. Appl. Phys.*, **51**, 06FJ07.
5. A. Fuchs, M. Bender, U. Plachetka, U. Hermanns, and H. Kurz (2005). Ultraviolet-based nanoimprint at reduced environmental pressure, *J. Vac. Sci. Technol. B*, **23**, 2925.
6. M. Berger and P. Colella (1989). Local adaptive mesh refinement for shock hydrodynamics, *J. Comput. Phys.*, **82**, 64.
7. H. Hiroshima and M. Komuro (2007). UV-nanoimprint with the assistance of gas condensation at atmospheric environmental pressure, *J. Vac. Sci. Technol. B*, **25**, 2333.

8. Y. Hirai, H. Kikuta, and T. Sanou (2003). Study on optical intensity distribution in photocuring nanoimprint lithography, *J. Vac. Sci. Technol.* B **21**, 2777.
9. H. Ichkawa and H. Kikuta (2001). Numerical feasibility study of the fabrication of subwavelength structure by mask lithography, *J. Opt. Soc. Am. A*, **18**, 1093.
10. R. Suzuki, N. Sakai, A. Sekiguchi, Y. Matsumoto, R. Tanaka, and Y. Hirai (2010).Evaluation of curing characteristics in UV-NIL resist, *J. Photopolym. Sci. Technol.*, **23**, 51.
11. R. Suzuki, N. Sakai, T. Ohsaki, A. Sekiguchi, H. Kawata, and Y. Hirai (2012). Study on curing characteristic of UV nanoimprint resist, *J. Photopolym. Sci. Technol.*, **25**, 211.
12. G. Odian (2004). *Principles of Polymerization*, John Wiley and Sons.

Chapter 8

Demolding Process Simulation

Yoshihiko Hirai
Department of Physics and Electronics Engineering,
Graduate School of Engineering, Osaka Prefecture University,
Gakuen-cho, Sakai, 599-8531 Osaka, Japan
hirai@pe.osakafu-u.ac.jp

8.1 Introduction

In nanoimprint lithography (NIL) [1, 2], one of the important issues is defect in the demolding process. To eliminate the defect, several approaches have been proposed, for example, by applying antisticking coatings on mold surfaces [3, 4], reducing friction between the mold and resin materials [5, 6], and using novel demolding methods [7, 8]. In addition to these approaches, many other experimental studies have been reported on the demolding process [9–14]. Nevertheless, few reports have been done on the mechanisms of the demolding process. Understanding the process mechanism is important to optimize the process and eliminate the defects.

In this chapter, numerical simulations are carried out in order to understand the demolding mechanisms from a mechanical point of view and confirmed by experiments [15, 16].

Nanoimprint and Its Applications
Edited by Akihiro Miyauchi

ISBN 978-981-4800-37-2 (Hardcover), 978-0-429-03192-2 (eBook)
www.jennystanford.com

8.2 Numerical Model [15]

8.2.1 Numerical Model

Numerical simulation is performed using the finite element method based on continuum mechanics using commercially available software MSC-marc [17]. An important model is the separation model between the mold and the resin to describe detachment.

A typical separation model is described on the basis of surface tension and considers the capillary force [18]. However, this model only considers normal loading to the boundary and neglects shear loading. To consider the shear force along the resin/mold boundary, we apply the virtual crack closure technique (VCCT) [19], in which the energy release rate near the crack tip is derived from the normal displacement and force. The VCCT can be used to simulate the separation of layers and crack propagation. This method has been successfully applied to analyze fracture phenomena.

A problem with this method is that parameter extraction of the stress intensity factors for three modes, the opening, sliding, and tearing modes, is experimentally challenging in the actual NIL process.

To solve the problem, a simplified model proposed by Ortiz and Pandolfi [20] is applied. On the basis of the model, the effective traction t is expressed as:

$$t = \sqrt{{t_n}^2 + t_s^2/\beta^2}\,, \tag{8.1}$$

where t_n and t_s are the nominal and shear traction of the cohesive layer, respectively [15]. The β expresses the contribution ratio between the shear and the normal critical traction. When the traction exceeds the critical opening force t_c, which is equivalent to the stored elastic strain energy exceeding the critical cracking energy, the cohesive layer opens and separation occurs.

Equation 8.1 reduces to

$$\sqrt{(t_n/t_c)^2 + (t_s/\beta t_c)^2} > 1\,. \tag{8.2}$$

The criterion for separation as expressed in Eq. 8.2 further reduces to

$$(\sigma_n/P_n)^2 + (\sigma_s/P_s)^2 > 1, \tag{8.3}$$

where σ_n and σ_s are the normal and shear stresses at the cohesive boundary and P_n and P_s are the "effective" critical normal and shear stress for separation, respectively. The separation criterion described by Eq. 8.3 at the boundary of the elements is described schematically in Fig. 8.1.

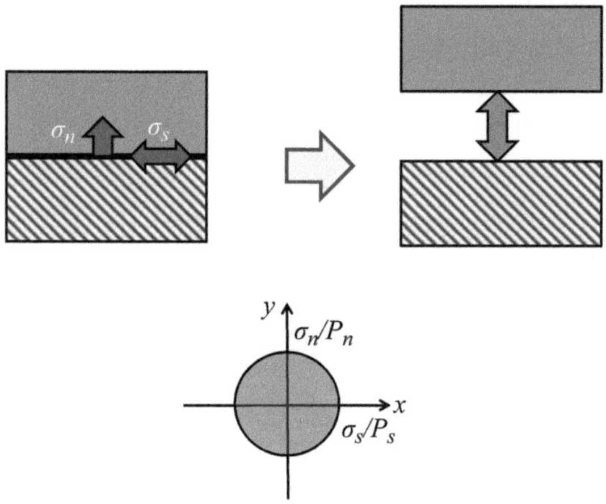

Figure 8.1 Diagram of the adhesive boundary criterion.

In the numerical calculation, a static point of the separation is calculated for each displacement of the mold and the reaction load of the mold is obtained for each displacement. Using the cohesive zone model, separation of the adhesive layers was simulated without crack initiation. The required critical normal and shear stress are easily obtained from experimental studies.

For friction between the resin and the mold, a simple conventional model was applied:

$$f = \eta \times f_n \tag{8.4}$$

where f_n is the normal force acting on the mold and η is a dynamic friction coefficient after separation at the stamp/resin interface occurs and motion of the interface begins.

8.2.2 Parameter Extraction by Experiment

P_n and P_s are important parameters for simulation. To verify these values, the values are experimentally evaluated [16]. Figure 8.2

shows the schematics of the verification method by scanning probe microscopy (SPM). We use a SiO_2 ball tip coated with an antisticking layer, OPTOOL DSX (Daikin Chemical) [3], to emulate a mold surface. The diameter of the SiO_2 ball is 0.35 mm. The tip is indented on a polymethylmethacrylate (PMMA) resin surface coated on the substrate. The critical normal stress P_n is extracted from the pull-off force from the resin per unit area, and the critical shear force P_s is extracted from the static friction force evaluated from the stick slip force in the lateral direction. The results are shown in Table 8.1.

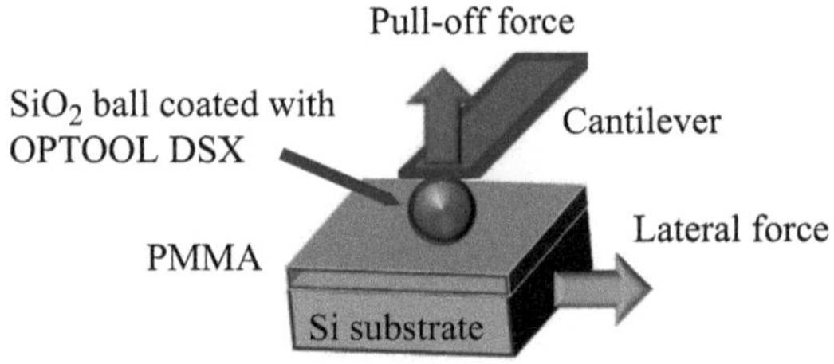

Figure 8.2 Schematics of the critical adhesive stress measurement using SPM. SPM, scanning probe microscopy.

Table 8.1 Experimental extraction of the critical stresses for Si mold/PMMA separation

Sample no.	1	2	3
P_n (kPa)	25.6	15.6	10.7
P_s (kPa)	180	313	195
P_s/P_n	7.0	20	18

On the other hand, we measured demolding loads for nonpatterned and patterned molds [21, 22]. For the PMMA/Si mold boundary, a critical normal stress P_n was reported around 5 kPa with an antisticking layer. For the patterned mold, a critical shear stress P_s was derived from the demolding load to be around 500 kPa. Those values are roughly consistent with each other.

The static friction coefficient was also reported by Ocypa et al. [23] to be approximately 0.02 at a PMMA/SiO_2 interface with an antistick coating; we used this value in our simulation.

8.3 Demolding Mechanism [15]

To analyze the demolding process, the relation between load and displacement is investigated. The simulation model is shown in Fig. 8.3, where *w*, *h*, and *d* represent the line width, pattern height, and residual thickness, respectively.

The resin was completely filled in the mold cavity, and no residual stress remained due to resin forming or shrinkage during thermal or UV curing. The resin is PMMA and is assumed to be an elastic body. The modulus of the resin is 1 GPa, and the Poisson ratio is 0.499. The mold material was assumed to be rigid.

The resin was gulled on the mold, and a fixed boundary was applied between the resin and the substrate. The mold was pulled up, and the displacement of the mold was varied. The reaction force as a demolding load was obtained as the loading force. The reaction force and the induced strain were calculated for each position (displacement) of the mold.

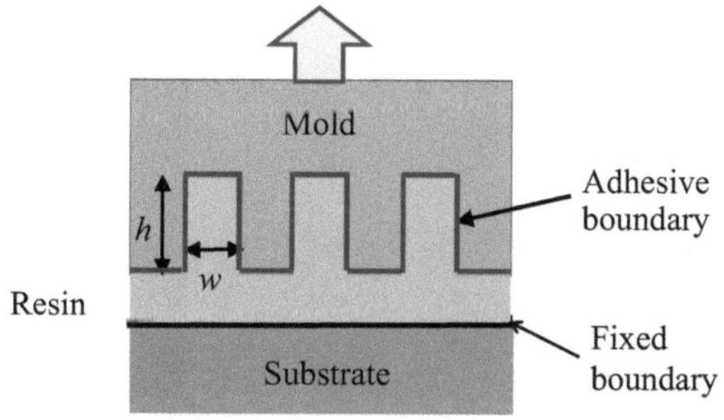

Figure 8.3 Schematic of the simulation model.

Figure 8.4 shows the simulation results of the demolding process. The process is expressed as several stages.

At first, the resin is stretched and elastic strain energy is accumulated both in the residual later and in the adhesive boundary. We assumed the critical normal stress P_n is 100 times larger than the critical shear stress P_s. The residual layer is firstly separated from the mold when the normal stress at the boundary exceeds a critical value because the stretching stress is concentrated in the residual region of the resin underneath the mold (Fig. 8.4b).

After separation of the residual layer, stress is concentrated along the sidewall boundary and sidewall separation begins (Fig. 8.4c) as the propagation of a crack. Then, the top portion of the resin separates (Fig. 8.4d). As a result, the mold is completely separated from the resin. Finally, the mold is lifted up as shown in Fig. 8.4e.

The upper views in Fig. 8.4 show normalized stress by elastic modulus (strain) distributions induced by mold lifting. Before separation of the residual layer (Fig. 8.4b), stress is concentrated in the corner of the resin pattern. The stress is released after separation of the residual layer and is concentrated at the tip of the crack during separation propagation along the sidewall (Fig. 8.4c). After separation, there is no induced stress (Fig. 8.4d). These results have, therefore, allowed us to comprehensively describe the release process.

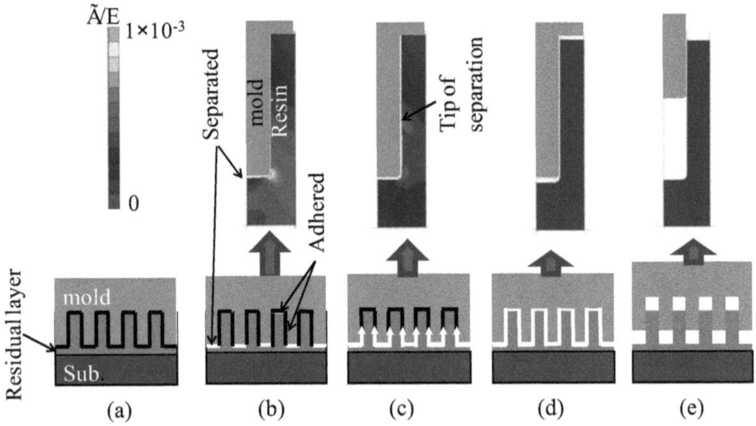

Figure 8.4 Schematics of the demolding mechanism and normalized principal stress (strain) distribution in the resin. (a) Initial state, (b) separation at the bottom, (c) separation along the sidewall, (d) complete separation, and (e) mold removal.

Figure 8.5 shows the simulated displacement-load curve, where Δd is displacement of the mold and d is the residual thickness. At the first stage, the load increases proportionately to the displacement because the elastic strain energy is accumulated in the resin and in the system (Fig. 8.5a). The demolding load reaches a maximum when the residual resin is separated from the mold (Fig. 8.5b). The load subsequently reduces and a small force is maintained during separation along the sidewall of the resin (Fig. 8.5c). After complete separation, the demolding load returns to zero and the mold is lifted off (Fig. 8.5d). When the sidewall is straight and the mold moves in the vertical direction, no normal force to the sidewall is induced and the dynamic friction force becomes zero.

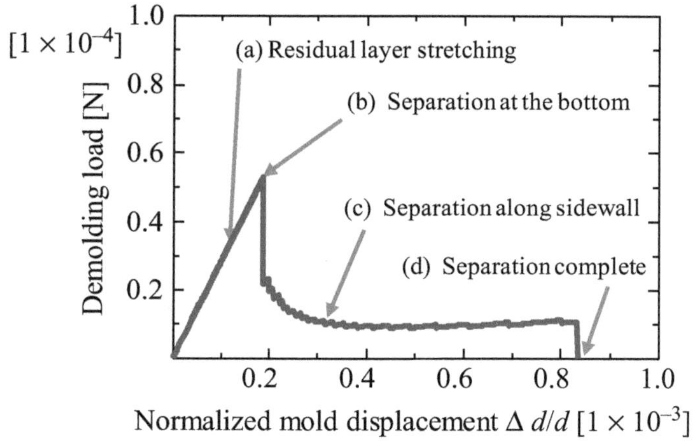

Figure 8.5 Simulated result of the demolding load versus displacement of the mold, where the displacement of the mold, Δd, is normalized by the residual thickness, d.

The mold releasing force (demolding load) is experimentally measured [22, 23]. The Si substrate is glued on the stage using epoxy bond, and the stage is lowered by the actuator using a stepping mortar. The demolding load is measured by a force gauge connected to the mold.

Typical displacement-load curves during demolding of PMMA from a Si mold with an antisticking coating are shown in Fig. 8.6, where the resin was PMMA with a typical thickness of around 5 μm and the mold was Si having a 2 μm line-and-space (L&S) pattern. At the starting stages, the load increases almost linearly and then decreases. The load then remains at a low level with small fluctuations. Finally, the load becomes zero and the mold is released. The maximum load required for release (i.e., the demolding load) is obtained from the first peak of the load. Other groups have also reported similar experimental results [10–12, 14]. The displacement force characteristic agrees well with the simulation in

Fig. 8.5, and the result is explained as follows: Firstly, elastic energy is accumulated in the resin (Fig. 8.5a). When the energy exceeds a critical value (Fig. 8.5b), separation starts at the residual layer and propagates along the sidewall (Fig. 8.5c). Then, the top plane lifts off and the separation process is finished (Fig. 8.5d). Finally, the mold is released (Fig. 8.5e).

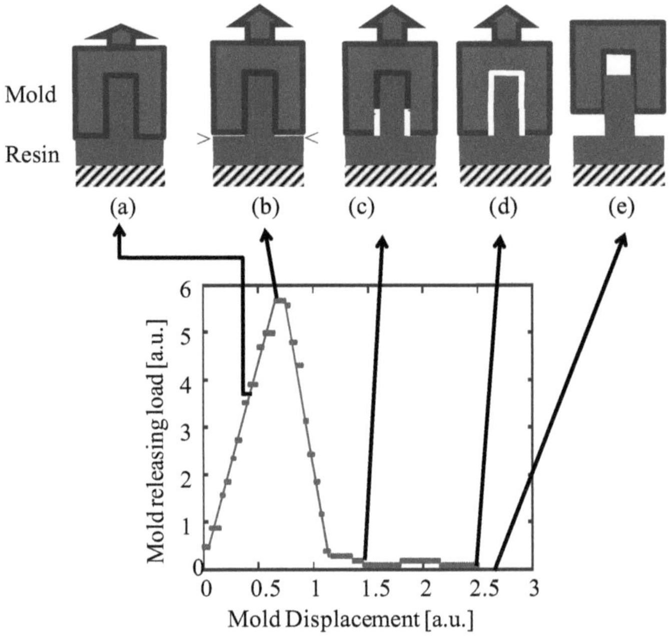

Figure 8.6 Experimentally determined demolding load. Typical result of the demolding load for a PMMA resin. The horizontal axis is the normalized displacement of the mold.

The simulation and experiment agree with each other, and the demolding mechanism is revealed.

8.4 Result and Discussion

8.4.1 Impact on the Critical Stresses P_s and P_n [16]

The demolding load strongly depends on the critical release stresses P_n and P_s, with low values favorable for release. In this section, the dependence of the demolding load on the sidewall quality factor P_s/P_n is investigated in detail by computational work. Figure 8.7a shows typical displacement force characteristics (force curve) in the demolding process when P_s/P_n is 1.

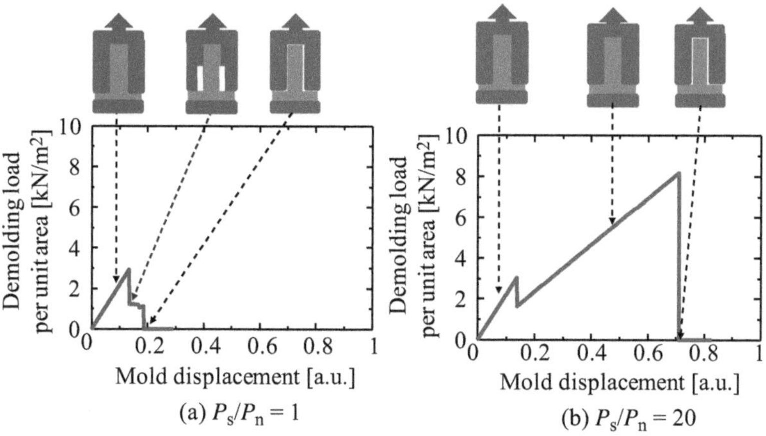

Figure 8.7 Simulation results of demolding force versus displacement of the mold.

The demolding load per unit area of the mold linearly increases by resin stretching. The stress induced at the bottom boundary of the mold to the residual layer is larger than that at the top boundary to the pattern cavity. Then, the bottom of the mold firstly separates from the residual layer and the demolding load decreases. As a result, the first peak appears in the demolding load curve. The first peak depends on the critical normal stress P_n because the interface between the resin and the bottom of the mold separates owing to the normal stress on the interface.

After the separation from the residual layer, the separation proceeds along the sidewall as the mold is pulled up. The adhering area decreases owing to the separation, and the demolding load remains nearly constant. Then, the sidewall of the mold separates from the resin and finally the upper side of the mold separates. After the separation, the demolding load decreases to zero. As a result, the demolding load curve shows multiple local peaks.

On the other hand, Fig. 8.7b shows a typical demolding load characteristic when P_s/P_n is 20. In this case, the critical shear stress P_s is larger than the normal critical stress P_n and the resin strongly sticks to the sidewall of the mold. The demolding load increases linearly owing to the elastic stretching of the resin until the sidewall is separated from the resin, and the second peak appears. Then, the separation of the resin from the sidewall immediately occurs under large shear stress and the top wall of the mold pattern is also separated. As a result, the demolding load decreases immediately to zero.

As discussed above, the maximum demolding load and the force characteristics depend on the P_s/P_n ratio, which restates the sidewall quality. When the P_s/P_n ratio increases, the demolding load has multiple peaks due to separation of the resin from the residual layer and sidewall of the pattern.

8.4.2 Impact on Aspect Ratio [15]

We examined the dependence of the demolding load on the aspect ratio of the pattern. Simulated force curves for various pattern aspect ratios are shown in Fig. 8.8a. The demolding load (i.e., the peak load) increases with increasing aspect ratio. The strain distributions in the resins are shown in Fig. 8.8b. The strain is distributed not only in the residual layer but also inside the cavities and is found to increase with increasing aspect ratio.

The simulated and experimental demolding loads for various aspect ratios are shown in Fig. 8.9. Both results show that the demolding load is approximately proportional to the aspect ratio to the power of 3/4. This is because the elastic strain energy is accumulated in the pattern cavity with increment of the aspect ratio. As a result, the demolding load increases. The simulation and the experimental results are in good agreement and suggest that the demolding load originates from the mold separation force from the residual layer of the resin.

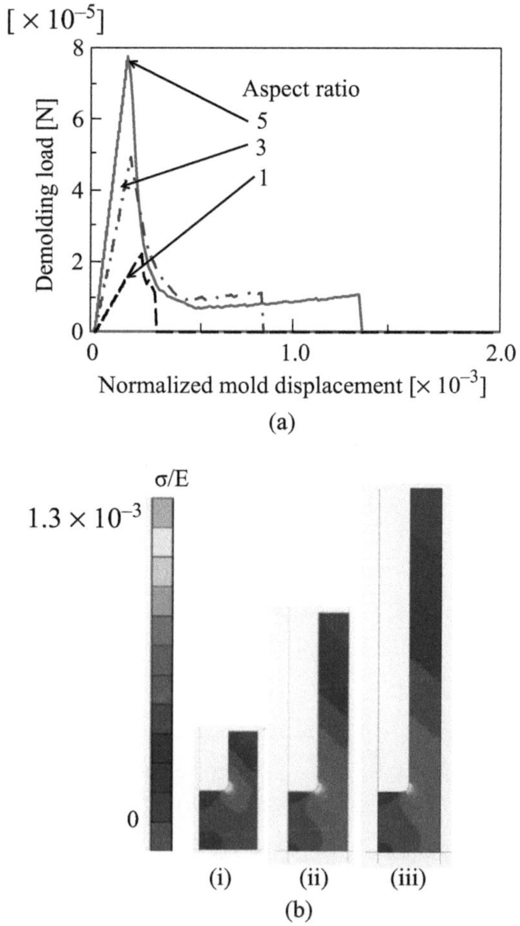

Figure 8.8 Dependence of the release load and induced strain on the aspect ratio of the pattern. (a) Demolding load characteristics and (b) induced principal strain distributions in various aspect ratios.

8.4.3 Impact on Sidewall Slope Angle [16]

To confirm the demolding mechanism, the impact of the sidewall slope angle θ is examined. Figure 8.10 shows schematics of the system. The sidewall slope angle θ is varied, and the demolding load is investigated. The pattern height is h, and the residual thickness of the resin is d.

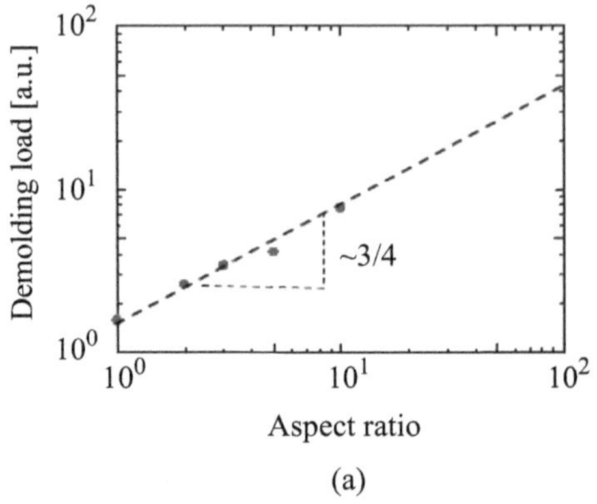

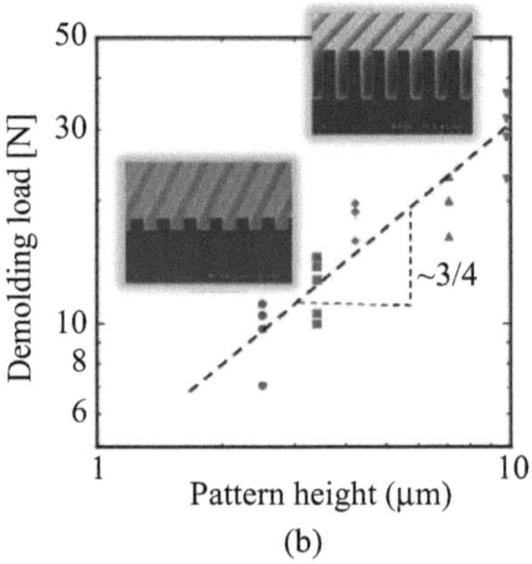

Figure 8.9 Dependence of demolding load on the aspect ratio of the pattern: (a) Simulation and (b) experimental results (width of the pattern = 2 μm).

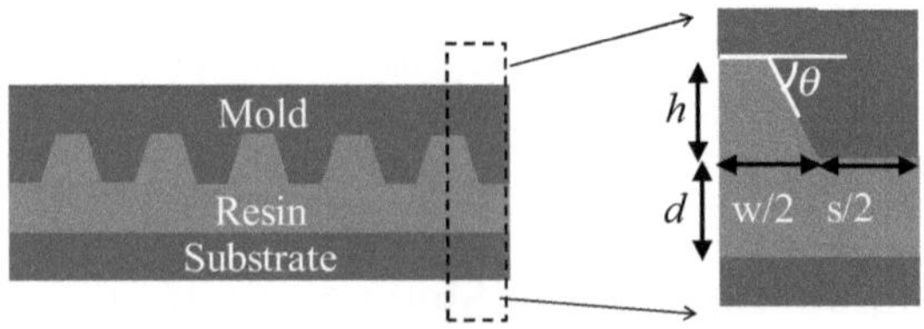

Figure 8.10 Schematic of the analytical model of resist separation from the mold. The pattern height is *h*, the residual layer is *d*, the pattern width is *w*, the space is *s*, and the sidewall slope angle is *θ*.

Figure 8.11 shows the simulation result of demolding load characteristics on the sidewall slope angle θ when P_s/P_n is 20. There arise multiple local peaks. The first peak shows mold separation from the residual layer and did not depend on the sidewall slope angles. The second peak is related to the sidewall separation from the resin, and it decreases with the decrement of the sidewall slope angle θ.

The first and second peaks become the same height at the critical sidewall slope angle θ_c, which depends on the pattern height, P_s/P_n ratio. When the sidewall slope angle exceeds the critical angle, the second local peak in the demolding load exceeds the first peak in the demolding load.

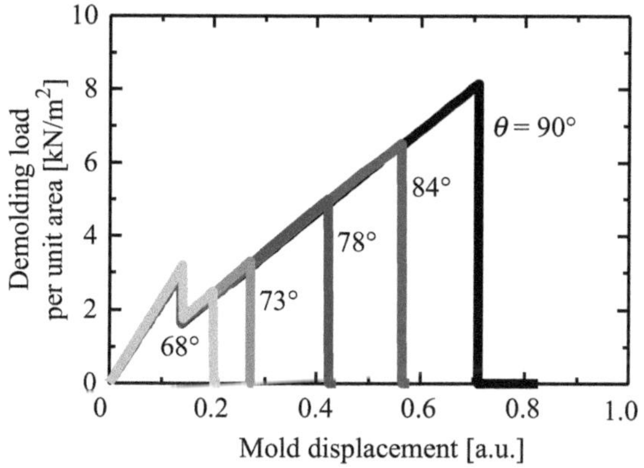

Figure 8.11 Computational results of the force curve at various sidewall slope angles θ at a ratio of $P_s/P_n = 20$.

Figure 8.12 shows the relation between the slope angle θ and the maximum demolding load at various P_s/P_n ratios. As the slope angle θ decreases, the demolding load decreases. Nevertheless, the release loads converge at a constant value when the θ is below the critical sidewall slope angle θ_c for any P_s/P_n ratio. The maximum demolding load depends on both the P_s/P_n ratio and the θ, which are related to the mold/resin boundary characteristics.

On the other hand, the demolding load is saturated when the P_s/P_n ratio becomes over 10. This is because a lateral force is induced in the resin pattern owing to vertical stretching and the normal stress σ_n at the interface on the sidewall exceeds the critical normal stress P_n before the shear stress σ_s exceeds the critical P_s shear stress [16].

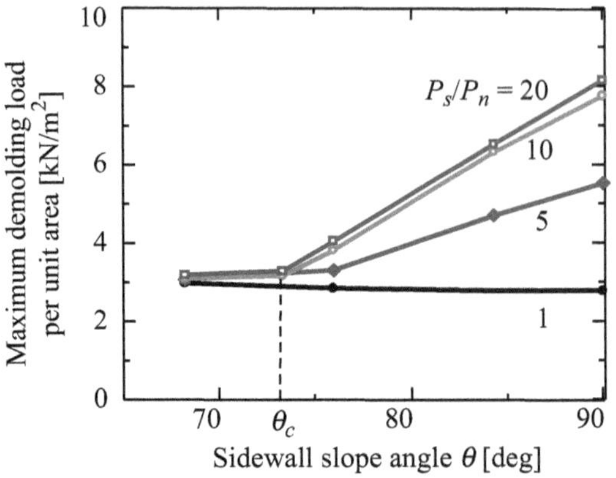

Figure 8.12 Computational results of the relationship between the sidewall slope angle θ and the maximum demolding load at various ratios.

To verify the simulation results, the demolding loads are experimentally measured using novel molds having various slope angles. The molds are prepared by conventional dry etching of a Si chip.

Figure 8.13 shows the cross-sectional views of the molds obtained by scanning electron microscopy. The pitch of the L&S pattern is 4 μm. The slope angle θ is varied from 90° to 63°. The pattern height is 2 μm. Using these molds, the demolding load is measured for PMMA, where the residual layers are all 2 μm thick.

Figure 8.14 shows typical results of the measured force curve in the demolding process for an angle θ = 80°. The result shows multiple peaks of the demolding load, which agrees well with the simulation result as shown in Fig. 8.11.

The impact of the critical stress and the slope angle of the mold on the demolding load is verified by computational works and compared with experiments.

Figure 8.15 shows the dependence of the maximum demolding load on the slope angle. The crosses indicate the measured demolding loads. The solid, dotted, and dashed lines indicate computational results at various P_s/P_n ratios. All of the demolding loads are normalized by the force at θ = 90° of the slope angle. The experimental results show that the force remains constant when the slope angle is decreased.

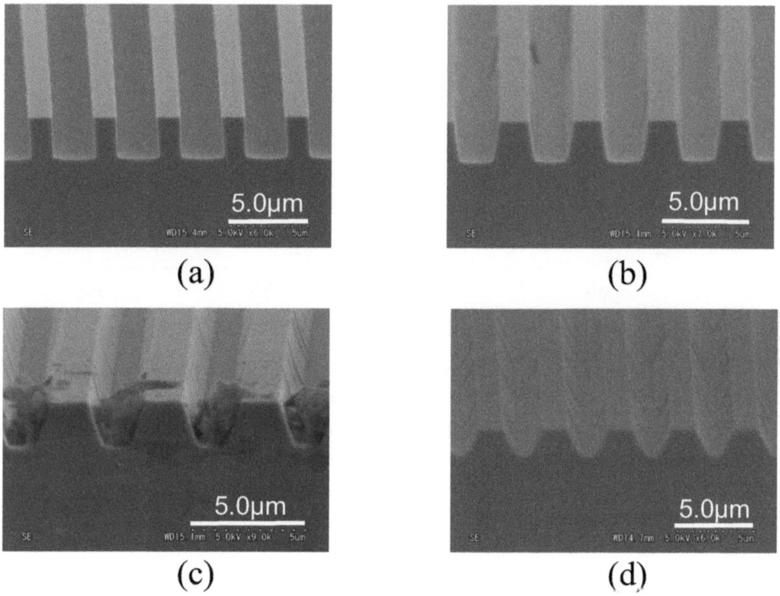

Figure 8.13 Cross-sectional views of the molds examined at various sidewall slope angles θ: (a) 90°, (b) 80°, (c) 70°, and (d) 62°.

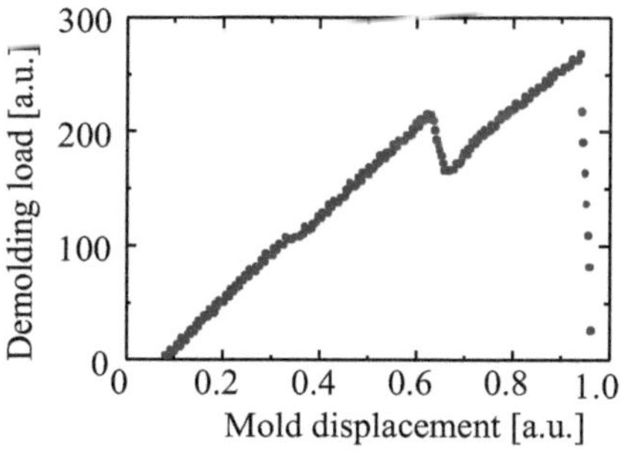

Figure 8.14 Example of the measured force curve in the mold release process. $\theta = 80°$.

The simulation and experimental results are in fairly good agreement for the impact of the sidewall slope angles.

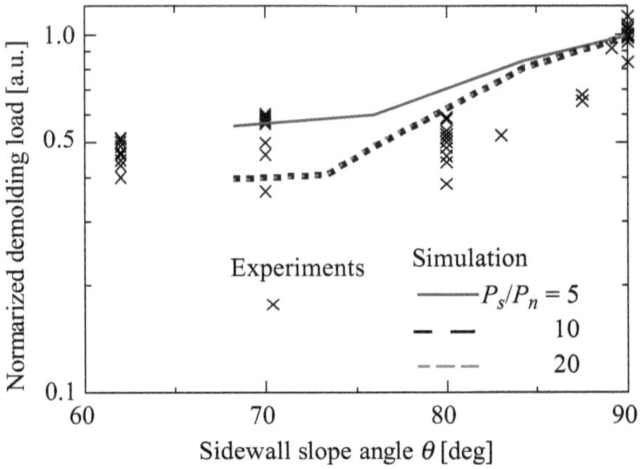

Figure 8.15 Simulation and experimental results of the maximum demolding load at various sidewall slope angles. The x marks indicate the measured demolding loads. The solid, dotted, and dashed lines indicate computational results. All of the demolding loads are normalized by the force at $\theta = 90°$.

8.5 Summary

The demolding process in NIL was simulated using a modified adhesive model based on fracture mechanics. In the first stage of the demolding process, the residual resin stretched and the mold separated from the residual resin layer. The separation ran subsequently along the sidewall of the mold, and the mold completely separated from the top of the resin layer. Following mold separation, the mold lifted up, thereby completing the release process.

The maximum demolding load was found to relate to the stretching of the resin before mold separation from the residual resin layer.

Experimental results also explain the impact on the aspect ratio and the sidewall slope angle of a mold, which quantitatively support the results of the simulation.

The computational results show that the shear separation force, which is related to the static friction between the mold surface and the resin, affects the demolding load characteristics when the ratio of the critical shear separation stress to the normal separation stress is below 10.

References

1. S. Y. Chou, P. Krauss, and P. Renstrom (1996). Nanoimprint lithography, *J. Vac. Sci. Technol.*, **B14**, 4129.
2. M. Colburn, S. Johnson, M. Stewart, S. Damle, T. Bailey, B. Choi, M. Wedlake, T. Michaelson, S. Sreenivasan, J. Ekerdt, and C. Willson (1999). Step and flash imprint lithography: a new approach to high-resolution patterning, *Proc. SPIE*, **3676**, 378.
3. Y. Hirai, S. Yoshida, A. Okamoto, Y. Tanaka, M. Endo, S. Lrie, H. Nakagawa, and M. Sasago (2001). Mold surface treatment for imprint lithography, *J. Photopolym. Sci. Technol.*, **14**, 457.
4. T. Bailey, B. Choi, M. Colburn, A. Grot, M. Meissl, S. Shaya, J. Ekerdt, S. Sreenivasan, and C. Willson (2000). Step and flash imprint lithography: template surface treatment and defect analysis, *J. Vac. Sci. Technol. B*, **18**, 3572.
5. B. Gates and G. Whitesides, Replication of vertical features smaller than 2 nm by soft lithography, *J. Chem. Soc.* **125** (2003)14986.
6. J. Gao, L. Yeo, M. Chan-Park, J. Miao, Y. Yan, J. Sun, Y. Lam, and C. Yue (2006). Antistick postpassivation of high-aspect ratio silicon molds fabricated by deep-reactive ion etching, *J. Microelectromech. Syst.*, **15**, 84.
7. H. Mekaru, O. Koizumi, A. Ueno, and M. Takahashi (2010). Inclination of mold pattern's sidewalls by combined technique with photolithography at defocus-positions and electroforming, *Microsyst. Technol.*, **16**, 1323.
8. T. Kitagawa, N. Nakamura, H. Kawata, and Y. Hirai (2014). A novel template-release method for low-defect nanoimprint lithography, *Microelectron. Eng.*, **123**, 65.
9. S. Park, Z. Song, L. Brumfield, A. Amirsadeghi, and J. Lee (2009). Demolding temperature in thermal nanoimprint lithography, *Appl. Phys. A*, **97**, 395.
10. V. Trabadelo, H. Schift, S. Merino, S. Bellini, and J. Gobrecht (2008). Measurement of demolding forces in full wafer thermal nanoimprint, *Microelectron. Eng.*, **85**, 907.
11. E. P. Chan and A. J. Crosbya (2006). Quantifying release in step-and-flash imprint lithography, *J. Vac. Sci. Technol. B*, **24**, 2716.
12. A. Amirsadeghi, J. J. Lee, and S. Park (2011). Surface adhesion and demolding force dependence on resist composition in ultraviolet nanoimprint lithography, *Appl. Surf. Sci.*, **258**, 1272.

13. A. Amirsadeghi, J. J. Le, and S. Park (2012). A simulation study on the effect of cross-linking agent concentration for defect tolerant demolding in UV nanoimprint lithography, *Langmuir*, **28**, 11546.

14. C. Hsueh, S. Lee, H. Lin, L. Chen, and W. Wang (2006). Analyses of mechanical failure in nanoimprint processes, *Mater. Sci. Eng. A*, **433**, 316.

15. T. Shiotsu, N. Nishikura, M.Yasuda, H. Kawata, and Y. Hirai (2013). Simulation study on the template release mechanism and damage estimation for various release methods in nanoimprint lithography, *J. Vac. Sci. Technol. B*, **31**, 06FB07.

16. T. Tochino, K. Uemura, M. Michalowski, K. Fujii, M. Yasuda, H. Kawata, Z. Rymuza, and Y. Hirai (2015). Computational study of the effect of side wall quality of the template on release force in nanoimprint lithography, *Jpn. J. Appl. Phys.*, **54**, 06FM06.

17. http://www.mscsoftware.com/product/marc

18. M. Park, Y. Lam, P. Laulia, and S. C. Joshi (2005). Simulation and investigation of factors affecting high aspect ratio UV embossing, *Langmuir*, **21**, 2000.

19. E. Rybicki and M. F. Kanninen (1997). A finite element calculation of stress intensity factors by a modified crack closure integral, *Eng. Fracture Mechanics*, **9**, 931.

20. M. Ortiz and A. Pandolfi (1999), Finite: deformation irreversible cohesive elements for three-dimensional crack-propagation analysis, *Int. J. Numer. Meth. Eng.*, **44**, 1267.

21. H. Kawata, M. Matsue, K. Kubo, M. Yasuda, and Y. Hirai (2009). Silicon template fabrication for imprint process with good demolding characteristics, *Microelectron. Eng.*, **86**, 700.

22. H. Kawata, J. Ishihara, T. Tanabe, M. Yasuda, and Y. Hirai (2012). A consideration of important factor on demolding force for various molds P223-6, *J. Photopolym. Sci. Technol.*, **25**, 222.

23. Ł. Ocypa, T. Nishino, H. Kawata, Y. Hirai, and Z. Rymuza (2012). Abstract of *Intl. Conf. Electron, Ion Photon Beam Technol.*, and *Nanofabrication* (May 2012, Waikoloa HW) P15–21.

Chapter 9

Measurement and Analysis Methods in Nanoimprinting

Makoto Okada[a] and Yoshihiko Hirai[b]

[a]*Laboratory of Advanced Science and Technology for Industry, University of Hyogo, 3-1-2 Koto, Kamigori, Ako, Hyogo 678-1205, Japan*

[b]*Graduate school of engineering, Osaka Prefecture University, 1-1 Gakuen-cho, Nakaku, Sakai, Osaka 599-8531, Japan*

m.okada@lasti.u-hyogo.ac.jp; hirai@pe.osakafu-u.ac.jp

Nanoimprinting technology can be used to rapidly and simply fabricate various types of nano/microstructures [1–9]. Scanning electron microscopy (SEM) and atomic force microscopy (AFM) are commonly used to observe the patterns produced by nanoimprinting technology because it is often difficult to confirm the small features of these patterns by optical microscopy. The authors describe principles of SEM and AFM observations in Sections 9.1 and 9.2. Frictional force measurements performed by AFM are also explained in Section 9.2. Nanoimprinting materials such as thermoplastic resins, ultraviolet (UV) curable resins, and sol-gel materials are important for nanoimprinting technologies because

Nanoimprint and Its Applications
Edited by Akihiro Miyauchi

ISBN 978-981-4800-37-2 (Hardcover), 978-0-429-03192-2 (eBook)
www.jennystanford.com

the required patterns depend on the properties of these materials. Furthermore, the release properties of the mold are important for performing nanoimprinting and the mold is typically coated with an antistick layer. However, molds without release properties are also used in some cases. The authors describe measurements of the glass transition temperature (T_g) and UV curing speeds of resins by differential scanning calorimetry (DSC) in Section 9.3. The calculation of the surface free energy by contact angle and rheology measurements of resins is explained in Sections 9.4 and 9.5.

9.1 Pattern Observation by Scanning Electron Microscopy

Scanning electron microscopy (SEM) is a commonly used technique for observing nanostructures [10]. Developments in SEM could lead to improvements in nanotechnology. In nanoimprinting experiments, the patterns of the mold and imprinted materials are typically confirmed by SEM.

Molds are usually fabricated by electron beam lithography (EBL) or photolithography. Silicon and quartz are frequently used as materials for the molds, in addition to metals such as Ni and Al. We are able to observe the patterns of Si and metal molds by SEM without the need for any treatment to prevent buildup of electrostatic charge. However, it is necessary to prevent electrostatic charging of the pattern for insulating materials such as quartz. The authors have observed patterns on a quartz mold without an antistatic treatment, as shown in Fig. 9.1. A JSM-6700F (JEOL Co.) SEM system was used in this section. Images were acquired at an accelerating voltage of 5 kV.

To obtain clear SEM images, a low-speed scan is typically used. However, images acquired by low-speed scanning of a quartz mold are blurred because of the electrostatic charging effects from the electron beam, as shown in Fig. 9.1a. We obtained clearer SEM images by scanning at a high-speed, as shown in Fig. 9.1b. However, it is better to perform an antistatic treatment to obtain clearer SEM images of insulating materials.

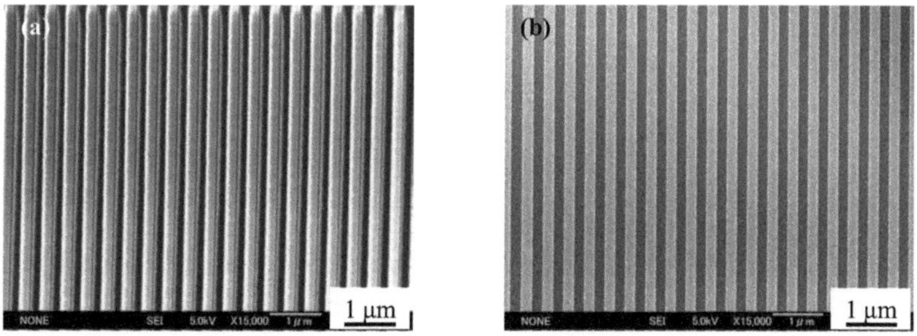

Figure 9.1 (a) Low-speed-scan and (b) high-speed-scan SEM images of quartz mold without treatment to prevent buildup of electrostatic charge. SEM, scanning electron microscopy.

To prevent electrostatic charging by the electron beam, a metal coating method is typically used. Figure 9.2a shows a low-speed-scan SEM image of a Au-coated quartz mold. The pattern of the mold could be clearly confirmed. However, cracking of the Au film can also be observed at high magnification, as shown in Fig. 9.2b. Furthermore, the thickness of the metal coating affects the pattern width and the metal is usually difficult to remove.

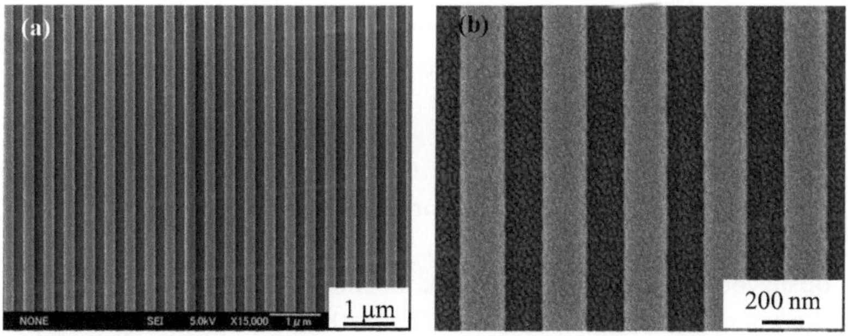

Figure 9.2 (a) Low-speed-scan SEM image of a Au-coated quartz mold. (b) High-magnification SEM image of the same pattern. SEM, scanning electron microscopy.

Another type of antistatic treatment involves the use of conductive polymer coatings. The authors used ESPACER (Showa Denko K. K.) as a conductive polymer, which is a charge dissipation polymer for EBL [11–13]. Incidentally, the quartz mold in this section was fabricated by EBL with the use of ESPACER. ESPACER is applied in the form of an aqueous solution, and the polymer can later be removed by washing with deionized water. The ESPACER was spin-coated onto the quartz mold, and we observed the pattern by SEM. Figure 9.3 shows a low-speed-scan SEM image of the ESPACER-coated quartz mold. The variation of contrast might be attributed to the ESPACER film thickness being insufficient to prevent electrostatic charging for low-speed scans. For measurements of the pattern widths, the film thickness might also affect the pattern widths. Hence, users need to consider the effects of any antistatic treatment.

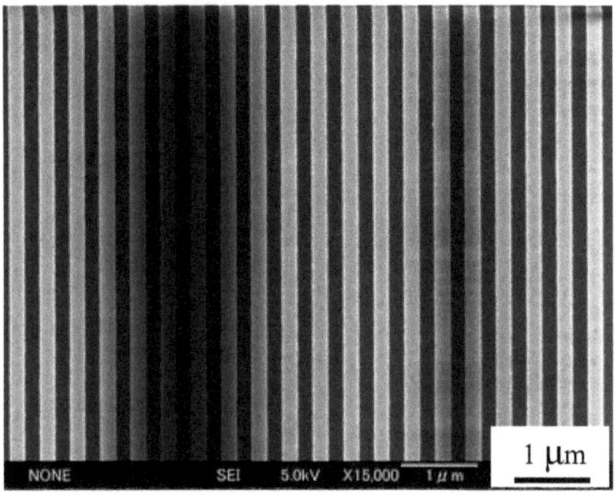

Figure 9.3 Low-speed-scan SEM image of an ESPACER-coated quartz mold. SEM, scanning electron microscopy.

It is also important to confirm the imprinted pattern on the replication material. The authors performed UV nanoimprinting using PAK-01 (Toyo Gosei Co.), which is a UV-curable resin, and observed the imprinted pattern, as shown in Fig. 9.4. In this case, the resin was coated with Au. The authors intentionally observed the central area in Fig. 9.4 for a long time and found that lines in this area deformed. This observation suggests that the curing reaction of PAK-01 proceeded further under the electron beam and the heat of electron beam affected the resin pattern. Thus, it is necessary to consider the effects of long-term electron beam irradiation on imprinted patterns in resin.

As described above, the authors used a 5 kV electron beam, and the electrical charge induced by the electron beam affected the SEM image. However, with the use of ultra-low-voltage SEM, the pattern could be observed without an antistatic treatment [14]. Therefore, users must select the appropriate SEM observation technique according to their experimental requirements.

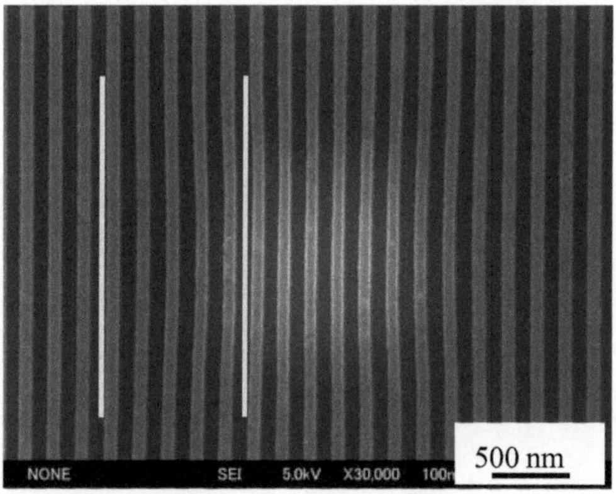

Figure 9.4 SEM image of PAK-01. Lines in the image are a guide to the eyes. The PAK-01 line near the right additional line was deformed. SEM, scanning electron microscopy.

9.2 Pattern Observation and Frictional Force Measurements by Atomic Force Microscopy

Scanning probe microscopy is a general term for several related technologies, which include scanning tunneling microscopy [15], atomic force microscopy (AFM) [16], Kelvin force microscopy [17, 18], and magnetic force microscopy [19, 20]. AFM is used to observe patterns. The tip of a cantilever comes in contact with the sample and scans its surface. As a result, the surface profile of the sample can be obtained. The cantilever is an important factor in AFM because the curvature radius of the cantilever affects the resolution of the AFM image. Generally, high-quality AFM images can be obtained with the use of a cantilever with a tip having a small curvature radius. AFM has three types of measurement modes: contact mode, tapping mode (dynamic force mode [DFM]), and noncontact mode. Users need to choose the appropriate mode for a sample. The authors used DFM-AFM to confirm the patterns in this section. The apparatus used was an E-sweep/NanoNavistation (Hitachi High-Tech Science Co.).

AFM images can be obtained without any antistatic treatment, such as metal coating, although it is better to perform electrical neutralization treatments because a cantilever might break through the static electrical charge of the sample, resulting in poor contact with the sample. Figure 9.5 shows AFM images of a pattern on a quartz mold without metal or conductive polymer coatings.

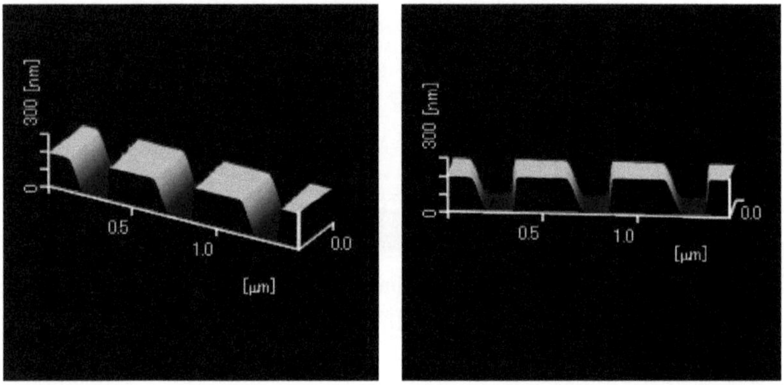

Figure 9.5 AFM images of an L&S pattern on a quartz mold. AFM, atomic force microscopy.

Clear 3D AFM images were obtained. We fabricated the cross-sectional surface by breaking the sample to observe a cross-sectional SEM image. However, the pattern in the AFM images shown in Fig. 9.5 does not fully reflect the actual rectangular pattern. Figure 9.6 illustrates AFM scanning of a rectangle pattern. The tip is tilted because of the AFM configuration used in the measurements described in this section (Fig. 9.6-1). The side of the tip comes in contact with the corner of the pattern (Fig. 9.6-2)] although the probe of the cantilever can scan the top of the pattern without any difficulty. Thus, the observed profile of the sidewall follows the dotted line, as shown in Fig. 9.6-5, which is not rectangular.

This problem becomes more pronounced for AFM observations of a high-aspect ratio pattern. Figure 9.7a shows an AFM image of a high-aspect-ratio pattern. The bottom of the image becomes sharp edged although the pattern is rounded, as shown by a cross-sectional SEM image of the same pattern (Fig. 9.7b). In addition, the pattern height of the AFM image, which was less than 2 μm, was lower than that shown by the SEM image, which was approximately 2 μm. Thus, the probe of the cantilever could not reach the bottom of the pattern in the AFM observations. Figure 9.8 shows an illustration of AFM scanning of a high-aspect-ratio pattern.

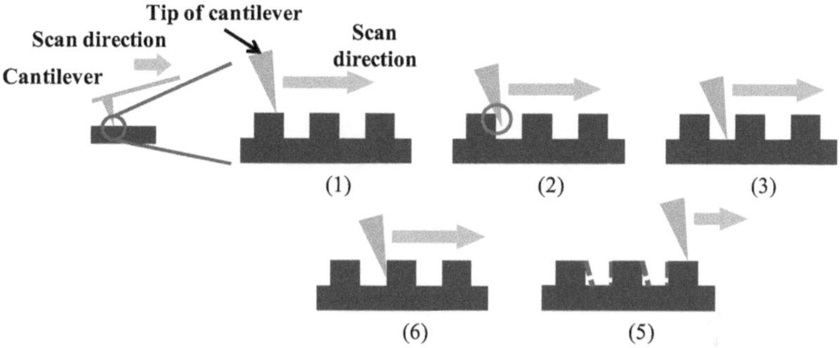

Figure 9.6 Illustration of AFM scanning on a rectangle pattern. AFM, atomic force microscopy.

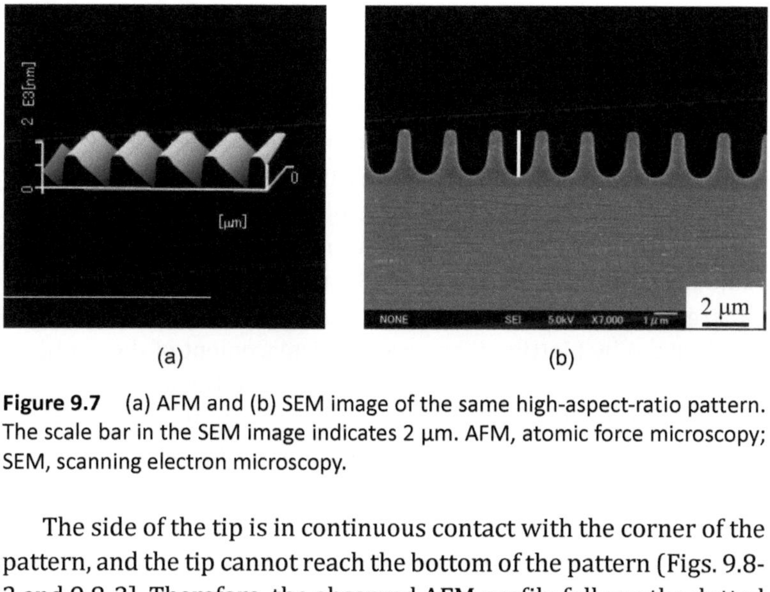

Figure 9.7 (a) AFM and (b) SEM image of the same high-aspect-ratio pattern. The scale bar in the SEM image indicates 2 μm. AFM, atomic force microscopy; SEM, scanning electron microscopy.

The side of the tip is in continuous contact with the corner of the pattern, and the tip cannot reach the bottom of the pattern (Figs. 9.8-2 and 9.8-3]. Therefore, the observed AFM profile follows the dotted line in Fig. 9.8-5.

As described above, AFM observations are useful for confirming pattern features without any special treatments. Furthermore, the height (or depth) of a pattern can be easily measured. However, users have to consider whether the AFM image is an accurate reflection of the sample without any cantilever issues. In addition, selection of the cantilever is important. For example, a high-aspect cantilever should be used to measure the high-aspect pattern.

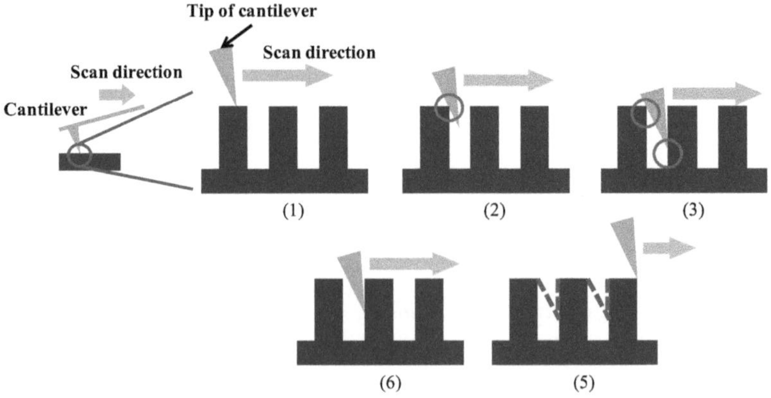

Figure 9.8 Illustration of AFM scanning of a high-aspect-ratio pattern. AFM, atomic force microscopy.

AFM can be used to measure the local frictional force on a surface because of the nanoscale cantilever probe [20–23]. Figure 9.9a shows an illustration of a frictional curve measurement by AFM. The torsional displacement of the cantilever is small if a low-friction surface is scanned because the cantilever can move smoothly on the surface. Conversely, a high-friction surface leads to a large torsional displacement of the cantilever. A frictional curve is obtained by plotting the torsional displacement of the cantilever via a reciprocating scan, as shown in Fig. 9.9b. The upper and lower sides indicate the measurement results from the approach and return routes. These results can be interpreted as the frictional force determined from the difference between the upper and lower sides of the frictional curve. Large and small differences between the upper and lower sides indicate high and low frictional forces, respectively. This frictional measurement is a relative measurement because the obtained frictional force is not an absolute value. Therefore, measurement of a reference sample is also needed to obtain quantitative results.

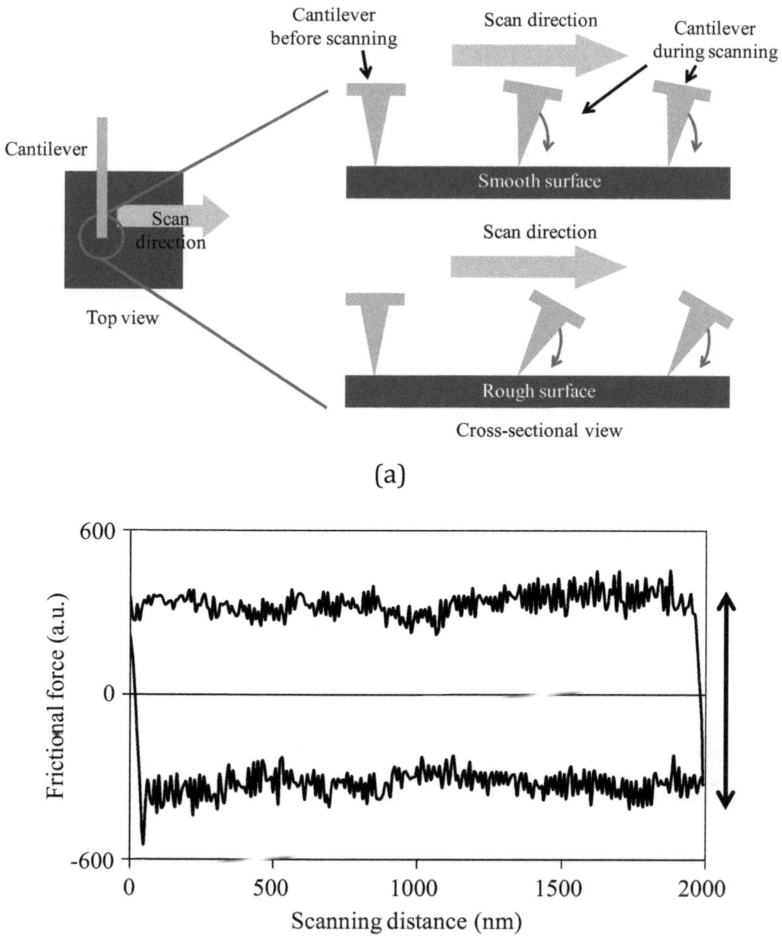

(a)

(b)

Figure 9.9 (a) Illustration of frictional curve measurement by AFM. (b) Example of a frictional curve. AFM, atomic force microscopy.

The authors measured the frictional force of fluorinated antistick layers. Generally, the nanoimprinting mold is coated with an antistick layer. Fluorinated self-assembled monolayers (F-SAMs) are typically used for this purpose. In these experiments a thermal oxide on a Si substrate (SiO_2/Si substrate) was used as the substrate. F-SAM was coated on the SiO_2/Si substrate with OPTOOL HD1100-TH (Daikin Industries). We measured the frictional curve on the SiO_2/Si and

F-SAM substrates. A spherical tip with a spring constant of 2.8 N/m (NanoWorld AG) was used as a cantilever. The curvature radius of the tip was 0.8 μm. The contact force of the cantilever, the scanning distance, and the scanning speed were approximately 200 nN, 2000 nm, and 1 Hz, respectively. Figures 9.10a and 9.10b show the frictional curves of the SiO_2/Si and F-SAM substrates, respectively. The frictional force of the F-SAM became considerable smaller than that of the SiO_2/Si substrate. These results clearly indicate that a low-friction surface could be obtained by forming the F-SAM.

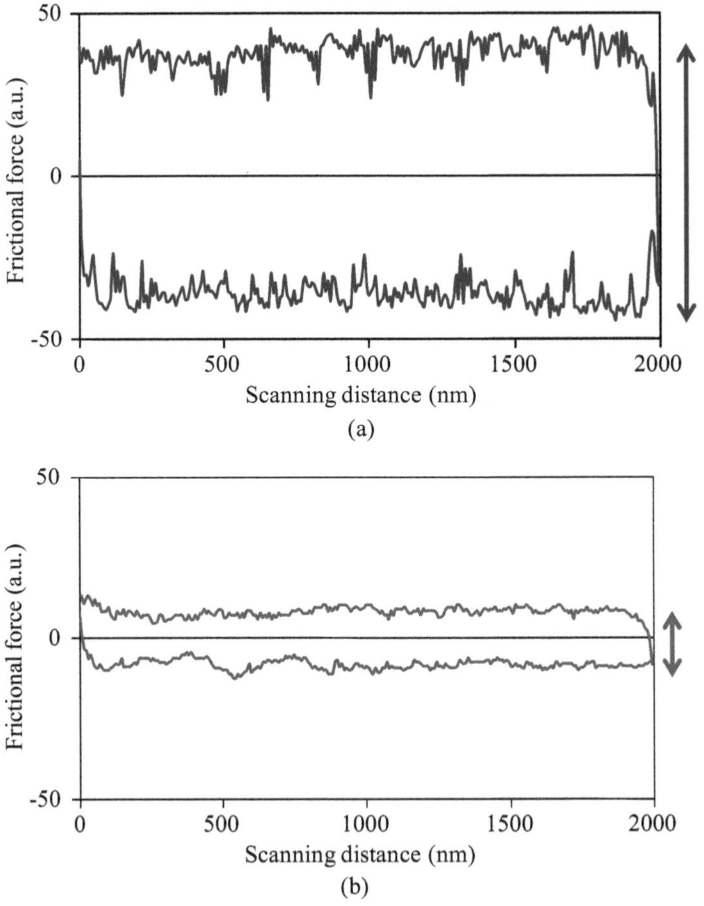

Figure 9.10 Frictional curves of (a) SiO_2/Si and (b) F-SAM substrates. F-SAM, fluorinated self-assembled monolayer.

9.3 Differential Scanning Calorimetry for Glass Transition Temperature and UV Curing Speed

The glass transition temperature (T_g) of a thermoplastic material is an important factor for performing thermal nanoimprinting because the thermal nanoimprinting temperature is typically higher than the T_g of the thermoplastic material. Additionally, the UV curing speed of a UV-curable resin is an indication of the UV irradiation time during UV nanoimprinting. In this section, the authors describe measurements of T_g and UV curing speed of materials by differential scanning calorimetry (DSC) [24].

The authors used functionalized polystyrene (PS) as a sample for T_g measurements. P4730-SOH is a commercially available hydroxyl-terminated PS (PS-OH). A PS layer is used in directed self-assembly of the block copolymer (BCP) [25–27]. One of the authors (Okada) has reported the chemical patterning of PS and polymethylmethacrylate (PMMA) layers fabricated by thermal nanoimprinting for BCP microphase separation [28]. According to the information of P4730-SOH provided from Polymer Source Inc., the T_g of P4730-SOH is 105°C. An EXSTAR DSC7010 (Hitachi High-Tech Science Co.) was used as the DSC apparatus. The heating rate was 10°C/min., and DSC scans were performed twice. Figure 9.11 shows the results of the second DSC cycle of PS-OH. The baseline was shifted to approximately 105°C. This phenomenon indicates a glass transition. The midpoint of the slope of the DSC result was approximately 105°C, which indicates a T_g.

Photo-DSC was applied to examine the curing speed of a UV-curable resin. PAK-01 (Toyo Gosei Co.) was used as a sample. The DSC apparatus was the EXSTAR DSC7010. The UV wavelength and intensity were 365 nm and 1 mW/cm^2, respectively, and the UV irradiation time was 20 min. Photo-DSC was performed twice on the same sample because the first DSC datum included the heat from the UV irradiation. The heat due to UV irradiation was determined by measuring the second photo-DSC after the UV-curable resin was perfectly cured by the first cycle of photo-DSC. To obtain the UV reaction heat of UV-curable resin, the second cycle of DSC data was deducted from first cycle of DSC data. Figure 9.12 shows the photo-DSC results of PAK-01. An exothermic peak occurred immediately after UV irradiation. The UV reaction time was estimated by measuring the time from the start of UV irradiation to the top of the exothermic peak.

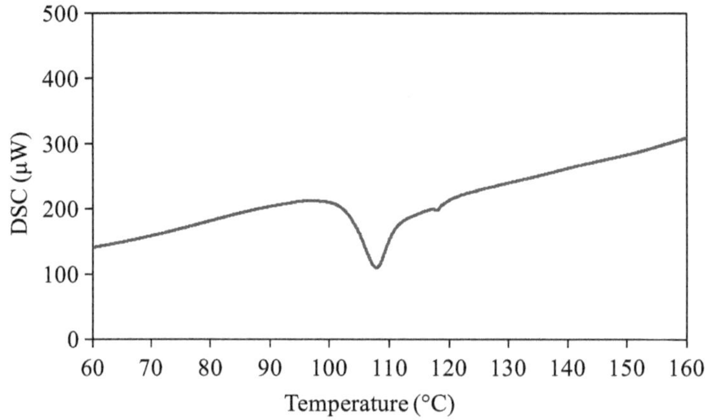

Figure 9.11 DSC results of PS-OH. DSC, differential scanning calorimetry; PS, polystyrene.

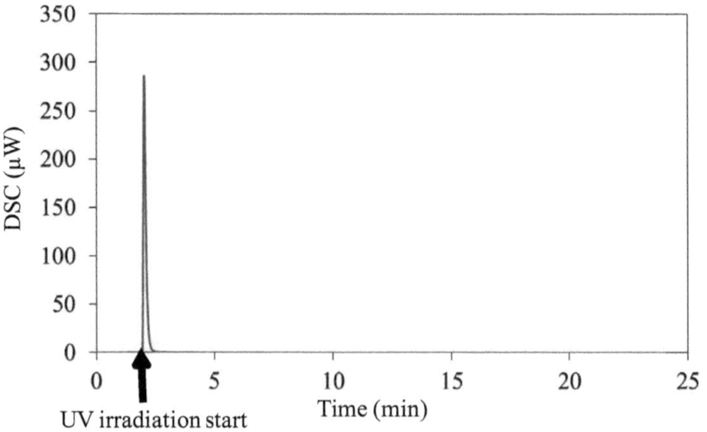

Figure 9.12 Photo-DSC result of PAK-01. DSC, differential scanning calorimetry.

References

1. S. Y. Chou, P. R. Krauss, and P. J. Renstrom (1995). Imprint of sub-25 nm vias and trenches in polymers, *Appl. Phys. Lett.*, **67**, 3114–3116.
2. S. Y. Chou, P. R. Krauss, and P. J. Renstrom (1996). Imprint lithography with 25-nanometer resolution, *Science*, **272**, 85–87.
3. S. Y. Chou, P. R. Krauss, W. Zhang, L. Guo, and L. Zhuang (1997). Sub-10 nm imprint lithography and applications, *J. Vac. Sci. Technol. B*, **15**, 2897– 2904.
4. J. Haisma, M. Verheijen, K. V. D. Heuvel, and J. V. D. Berg (1996). Mold-assisted nanolithography: A process for reliable pattern replication, *J. Vac. Sci. Technol. B*, **14**, 4124–4128.
5. T. Bailey, B. J. Choi, M. Colburn, M. Meissi, S. Shaya, J. G. Ekerdt, S. V. Sreenivasan and C. G. Willson (2000). Step and flash imprint lithography: Template surface treatment and defect analysis, *J. Vac. Sci. Technol. B*, **18**, 3572–3577.
6. S. Matsui, Y. Igaku, H. Ishigaki, J. Fujita, N. Ishida, Y. Ochiai, M. Komuro, and H. Hiroshima (2001). Room temperature replication in spin on glass by nanoimprint technology, *J. Vac. Sci. Technol. B*, **19**, 2801–2805.
7. F. Hua, Y. Sun, A. Gaur, M. A. Meitl, L. Bilhaut, L. Rotkina, J. Wang, P. Geil, M. Shim, J. A. Rogers, and A. Shim (2004). Polymer imprint lithography with molecular-scale resolution, *Nano Lett.*, **14**, 2467–2471.
8. Y. Miyake, Y. Akita, H. Oi, M. Mita, S. Kaneko, K. Koyama, K. Sunagawa, K. Tada, Y. Hirai, and M. Yoshimoto (2011). Nanoimprint fabrication and thermal behavior of atomically ultrasmooth glass substrates with 0.2-nm-height steps, *Jpn. J. Appl. Phys.*, **50**, 078002 1–2.
9. T. Kono, M. Hatano, H. Tokue, K. Kobayashi, M. Suzuki, K. Fukuhara, M. Asano, T. Nakasugi, E. H. Choi, and W. Jung (2017). Study of nanoimprint lithography (NIL) for HVM of memory devices, *Proc. SPIE*, **10144**, 1014406.
10. NTS Inc. (2008). Nano Imaging, (In Japanese), etc.
11. Showa Denko K. K. website, http://www.sdk.co.jp/products/45/75/1292/detail.html (in Japanese), http://www.sdk.co.jp/english/news/2005/aanw_05_0332.html, (accessed October 23, 2017).
12. S. Omoto, M. Okada, Y. Kang, K. Kanda, Y. Haruyama, S. Tono, and S. Matsui (2010). Comparison of EB exposure characteristics between HSQ and calix arene of high resolution negative resist, *J. Photopolym. Sci. Technol.*, **23**, 97–100.

13. D. A. Westly, D. M. Tennant, Y. Aida, H. Ohki, and T. Ohkubo (2011). Improved time dependent performance of hydrogen silsesquioxane resist using a spin on top coat, *J. Vac. Sci. Technol. B*, **29**, 06FJ02.
14. JEOL NEWS Volume 48 (2013). Keysight Technologies, Application Note, Low voltage scanning electron microscopy promises and challenges, (2012).
15. G. Binning, H. Rohrer, Ch. Gerber, and E. Weibel (1982). Surface studies by scanning tunneling microscopy, *Phys. Rev. Lett.*, **49**, 57–61.
16. G. Binning, C. F. Quate, and Ch. Gerber (1986). Atomic force microscope, *Phys. Rev. Lett.*, **56**, 930–933.
17. M. Nonnenmacher, M. P. O'Boyle, and H. K. Wickramasinghe (1991). Kelvin probe force microscopy, *Appl. Phys. Lett.*, **58**, 2921.
18. J. M. R. Weaver and D. W. Abraham (1991). High resolution atomic force microscopy potentiometry, *J. Vac. Sci. Technol. B*, **9**, 1559.
19. D. Rugar, H. J. Mamin, P. Guethner, S. E. Lambert, J. E. Stern, I. McFadyen, and T. Yogi (1990). Magnetic force microscopy: General principles and application to longitudinal recording media, *J. Appl. Phys.*, **68**, 1169.
20. B. Bhushan, H. Fuchs, and H. Hosaka (Eds.) (2004). *Applied Scanning Probe Methods*, Springer-Verlag, Berlin, Heidelberg.
21. C. M. Mate, G. M. McClelland, R. Erlandsson, and S. Chiang (1987). Atomic-scale friction of a tungsten tip on a graphite surface, *Phys. Rev. Lett.*, **59**, 1942–1945.
22. R. Erlandsson, G. Hadziioannou, C. M. Mate, G. M. McClelland, and S. Chiang (1988). Atomic scale friction between the muscovite mica cleavage plane and a tungsten tip, *J. Chem. Phys.*, **89**, 5190–5193.
23. B. Bhushan (Ed.) (2005). *Nanotribology and Nanomechanics*, Springer.
24. The Japan Society of Calorimetry and Thermal Analysis (Ed.) (2010). *Handbook of Calorimetry and Thermal Analysis*, Maruzen (in Japanese)
25. R. A. Segalman, H. Yokoyama, and E. J. Kramer (2001). Graphoepitaxy of spherical domain block copolymer films, *Adv. Mater.*, **13**, 1152–1155.
26. J. Y. Cheng, C. A. Ross, E. L. Thomas, H. I. Smith, and G. J. Vancso (2002). Fabrication of nanostructures with long-range order using block copolymer lithography, *Appl. Phys. Lett.*, **81**, 3657–3659.
27. S. O. Kim, H. H. Solak, M. P. Stoykovich, N. J. Ferrier, J. J. Pablo, and P. F. Nealey (2003). Epitaxial self-assembly of block copolymers on lithographically defined nanopatterned substrates, *Nature*, **424**, 411–414.
28. H. Wakaba, M. Okada, S. Iyoshi, Y. Haruyama, and S. Matsui (2014). Chemical patterns fabricated by nanoimprinting for block copolymer directed self-assembly, *J. Photopolym. Sci. Technol.*, **27**, 99–102.

Chapter 10

Applications

Akihiro Miyauchi[a] **and Hiroshi Sato**[b]
[a]*Tokyo Medical and Dental University, 2-3-10 Kanda-Surugadai, Chiyodaku, Tokyo 101-0062, Japan*
[b]*Canon Inc., 20-2, Kiyohara-Kogyodanchi, Utsunomiya-shi, Tochigi 321-3292, Japan*
miyauchi.ibb@tmd.ac.jp, sato.hiroshi092@canon.co.jp

Applications of nanoimprinting are rapidly expanding into various industrial fields. In this chapter, applications in the optics, biology, energy, and electronics fields are introduced.

10.1 Optical Applications

As shown in Fig. 1.3, the fabrication size by nanoimprinting ranges from a single nanometer to submillimeters. Nanoimprinting is, therefore, suitable for fabricating optical devices because the fabrication size of nanoimprinting covers the wavelength of visible light (i.e., 400–700 nm).

Several functions of optical components fabricated by nanoimprinting have been reported [1–10]. From an industrial viewpoint, displays are attractive application fields for

Nanoimprint and Its Applications
Edited by Akihiro Miyauchi

ISBN 978-981-4800-37-2 (Hardcover), 978-0-429-03192-2 (eBook)
www.jennystanford.com

nanoimprinting because liquid-crystal display (LCD) panels have many optical components, such as light guides and light diffusers, color filters, and polarizers. Feature sizes of these components are several dozen nanometers to several microns. Advances in displays have increased the demand for higher resolution, thinner panels, and lower power consumption. As a result, optical components have become one of the leading application fields of nanoimprinting.

10.1.1 Antireflection

When mobile phones or LCDs are viewed, the incident light on the display is reflected at the display surface and visibility is degraded. Therefore, an antireflection coating is usually applied to the surface of optical displays to improve their visibility. A conventional antireflection coating functions according to Bragg's law, and multilayered films are used as an antireflection coating (see Appendix D for Bragg's law).

A certain kind of 3D structure, namely a moth eye, has an antireflection property. As shown in Fig. 10.1, a moth eye has a lot of tiny protrusions, several hundred nanometers in size. The principle of antireflection of a moth eye structure is shown in Fig. 10.2. When incident light enters the protrusions, because the size of the protrusions is less than the wavelength of the incident light, the light does not get diffracted. Under the assumption that the protrusions are quadrangular pyramids with the same refractive index as that of the substrate, the refractive index is gradually increased until it reaches the refractive index of the substrate, as shown in Fig. 10.2b. According to the Fresnel equation, reflection occurs at an optical interface, namely an interface between two mediums with different refractive indices. In the moth eye structure, no optical interfaces exist, so reflection does not occur.

The reflectance characteristic of the moth eye structure was investigated as follows. The structure was formed on a spherical lens. Soft mold technology (described in Section 4.3.2) was applied to imprint the structure on a curved surface. Reflectance of lenses with and without the moth eye structure was plotted in Fig. 10.3. Clearly, reflectance is reduced by about 1/7 in the visible light range. This considerable reduction of reflectance confirms that nanoimprinting is useful for obtaining an antireflection effect. An antireflection film with a moth eye structure is already in mass production using roll-type photonanoimprinting.

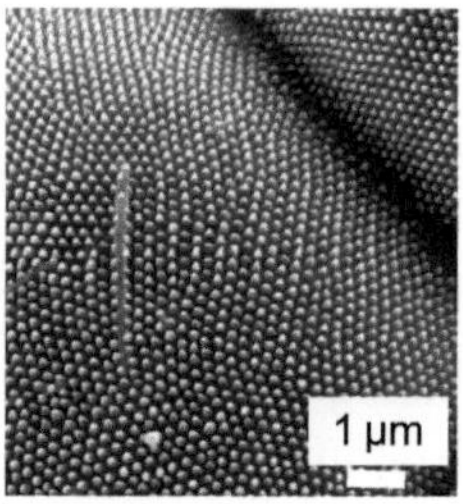

Figure 10.1 SEM image of a moth eye. SEM, scanning electron microscopy.

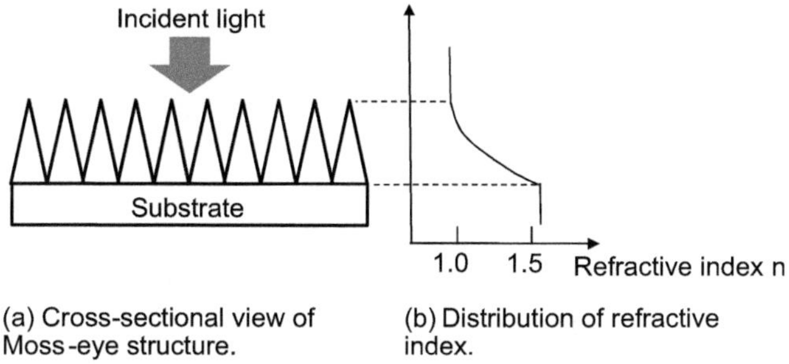

Figure 10.2 (a) Schematic diagram of cross-sectional view of a moth eye structure and (b) distribution of refractive index.

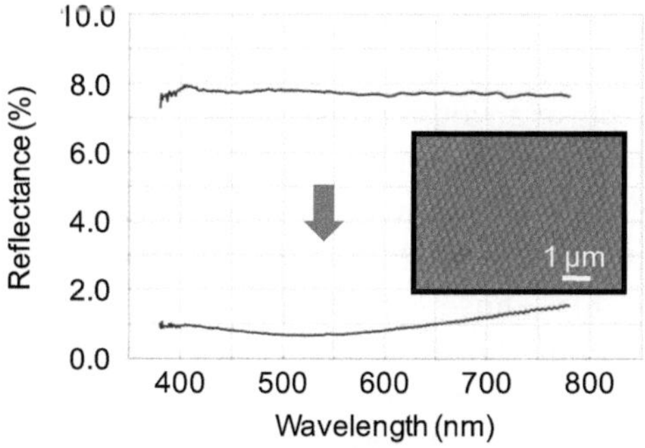

Figure 10.3 Reflectance of lens with moth eye structure as a function of incident wavelength. The reflectance is drastically reduced around 1/7 by the moth eye. The moth eye structure was formed on the surface of the spherical lens by soft mold technology.

10.1.2 Polarizer

Polarization of light is essential for many optical devices, especially LCDs and liquid-crystal projectors. Fabrication of a polarizer by nanoimprinting has been reported [8, 9]. As an example of photonanoimprinting, the fabrication process for a wire grid polarizer is described in Section 4.5. Here, the principle of a wire grid polarizer is explained.

An LCD projector using polarizers is shown schematically in Fig. 10.4. White light is split into light of three wavelengths: blue, green, and red. Two polarizers are set in the path of each light, and a liquid-crystal panel is inserted between each set of polarizers to control the polarizing direction.

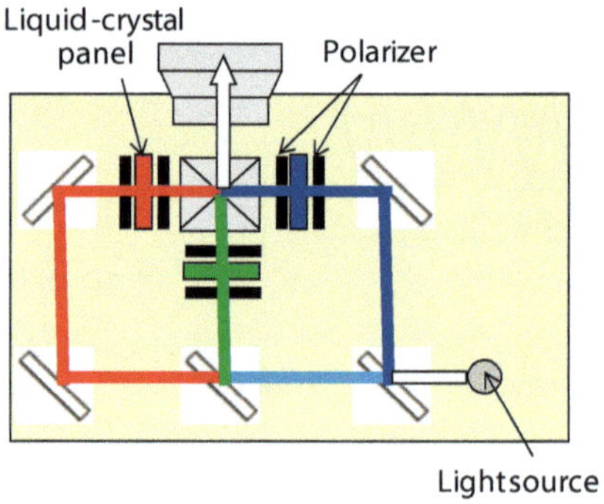

Figure 10.4 Diagram of LCD projector. The white light is split into three colors, namely blue, green, and red. Two polarizers are set in each light path. LCD, liquid-crystal display.

The principle of an aluminum-wire-grid polarizer is explained in Fig. 10.5. The aluminum wires are fabricated on a glass substrate. The width of the aluminum wires depends on the wavelength and is about 50 nm for blue light. When light enters the polarizer, only p-polarized light (whose electric field is parallel to the aluminum wire) can penetrate the polarizer, because the scattering wave of the electric dipole occurs only in the direction of the wires.

Polarized images with fine and defective patterns are shown in Fig. 10.6. In the case of the fine imprinted pattern, uniform bright and dark images are observed in the transmission image of p-polarized light (T_p) and the transmission image of s-polarized light

(T_S), respectively. However, in the case of the defective imprinted pattern, dark spots appear in the p-polarized image and bright stripes appear in the s-polarized image. These irregular regions are caused by nanoimprinting defects such as resist peeling. A means of reducing the occurrence of such defects, by improving materials and processes, is described in Section 4.5.

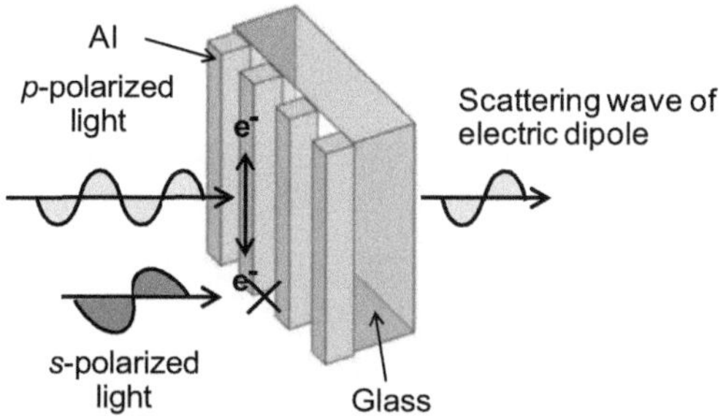

Figure 10.5 Principle of polarization by aluminum wire grid.

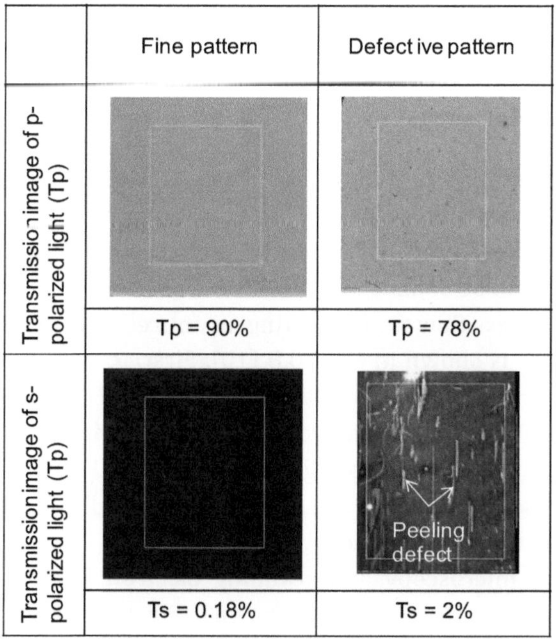

Figure 10.6 Polarized images of p-wave and s-wave.

The fine and defective nanoimprinted polarizers were installed in an LCD projector, and projected images were evaluated. The projector's characteristics are shown in Fig. 10.7. It is clear from the figure that a large defect generates the transmission of light in the s-polarized image and causes irregular spots in the white and blue images. In contrast, fine color images were obtained by using fine imprinted polarizers.

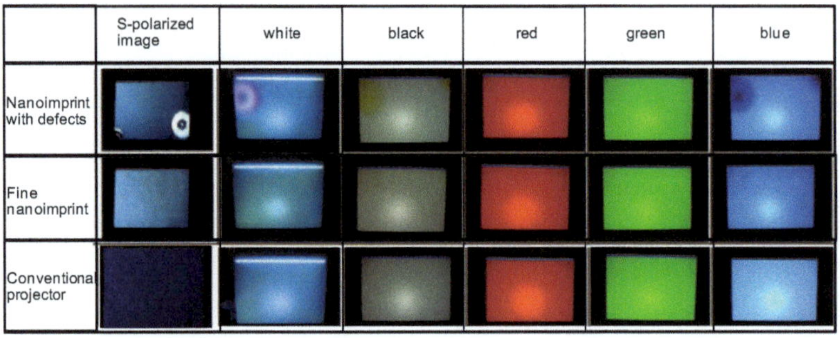

Figure 10.7 Projector characteristics of assembled nanoimprint polarizer. The nanoimprint polarizer is set in a blue light path.

10.1.3 Microlens

A microlens is an important component of optical sensors, LEDs, cameras, and other electronic devices. By using the antireflection technology described in Section 10.1.1, the optical characteristic of a microlens can be enhanced.

The process flow for fabricating an antireflection layer on top of a microlens is shown in Fig. 10.8 [10]. First, to make a template, a foil with an antireflection pattern and a mold for a lens array are prepared. Then, the template for the microlens with an antireflection layer is created by integrating the foil on the microlens mold. This template is used to create the replica mold. Finally, the microlens is molded by the replica mold.

Optical microscopy and scanning electron microscopy (SEM) images of the fabricated lens are shown in Fig. 10.9. The antireflection (moth eye) structure is formed on the lens surface. Transmission through microlenses with and without the antireflection structure is plotted in Fig. 10.10. It is clear from the figure that transmission is enhanced by the antireflection structure.

A hybrid composed of nano- and microstructures is attractive for adding new properties to optical devices like this microlens. The template for hybrid nano- and microstructures is created by integrating the two kinds of template. Therefore, the integration process (i.e., using a soft mold) of two kinds of templates is the key point.

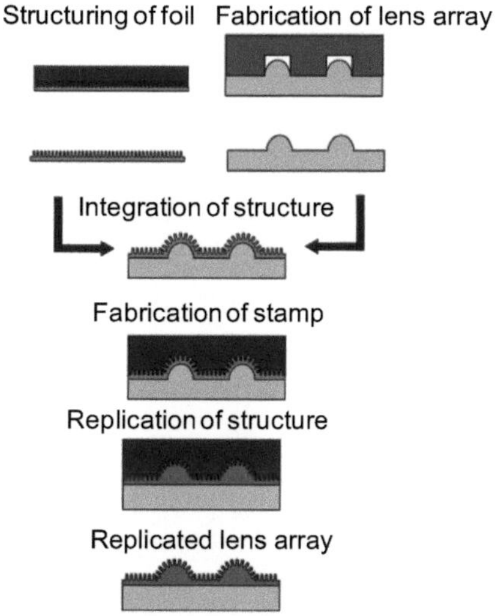

Figure 10.8 Process flow of microlens with antireflection structure. Reprinted with permission from Ref. [10]. Copyright 2011, American Vacuum Society.

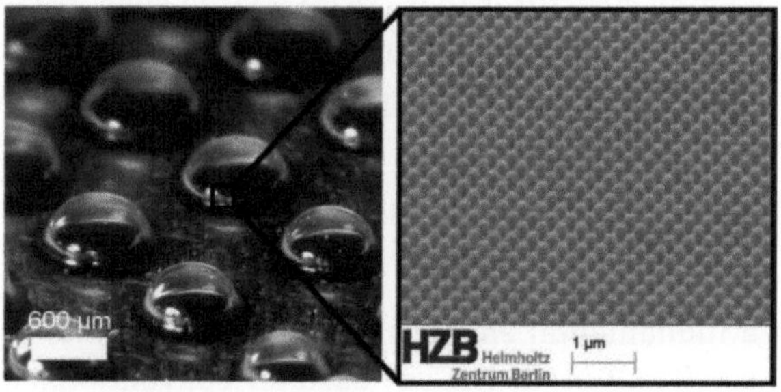

Figure 10.9 Fabricated microlens with antireflection structure. Reprinted with permission from Ref. [10]. Copyright 2011, American Vacuum Society.

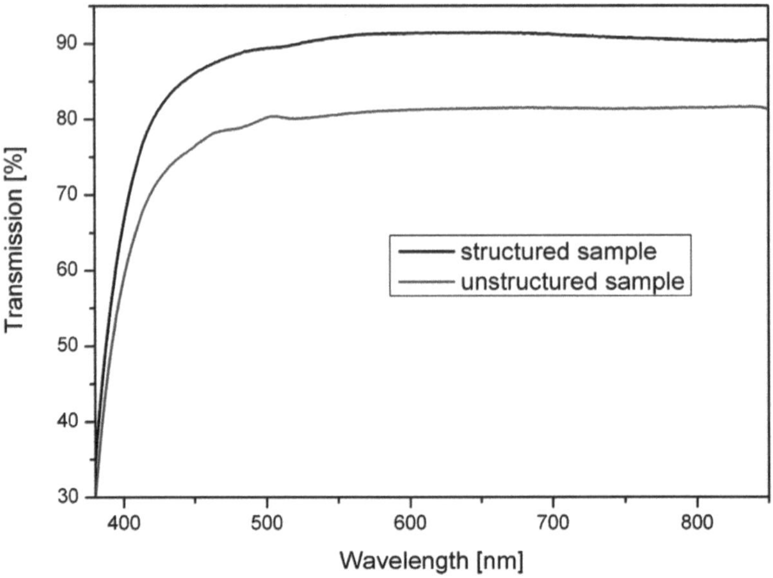

Figure 10.10 Transmission of microlenses with and without the antireflection structure. Reprinted with permission from Ref. [10]. Copyright 2011, American Vacuum Society.

10.2 Biodevices

The sizes of stretched DNA and cells are about several to several dozen microns. Therefore, nanoimprinted structures can be used to control these biomaterials directly [11, 12]. And a nanoimprinted surface has tiny structures that create a large specific surface area. These features are attractive in terms of biorelated devices. In this section, applications of nanoimprinting in biorelated devices are introduced.

10.2.1 Cell Culture

A cell is a fundamental element of living things. And, as shown in Fig. 10.11, cell culture dishes are used for conventional cell-growth experiments. Cell growths on a conventional flat-bottom dish and on a nanoimprinted "nanopillar scaffold" are shown in Figs. 10.12a and 10.12b, respectively [13, 14]. Both materials are polystyrene (PS), and the cell is a HeLa cell. HeLa cells stick in a flat manner on the conventional dish; however, on the nanoimprinted pillar structure, they become spherical shaped. The sticking in the former case is caused by an adhesive protein. Therefore, the difference in how a cell adheres to a scaffold when a flat dish is used and how it does so when a pillar surface is used was investigated.

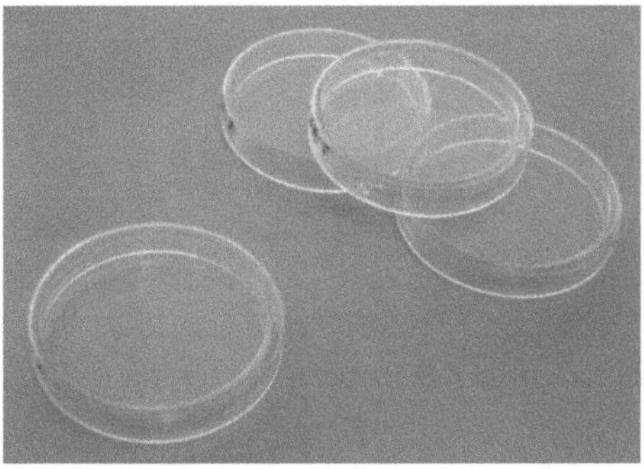

Figure 10.11 Appearance of conventional cell culture dishes.

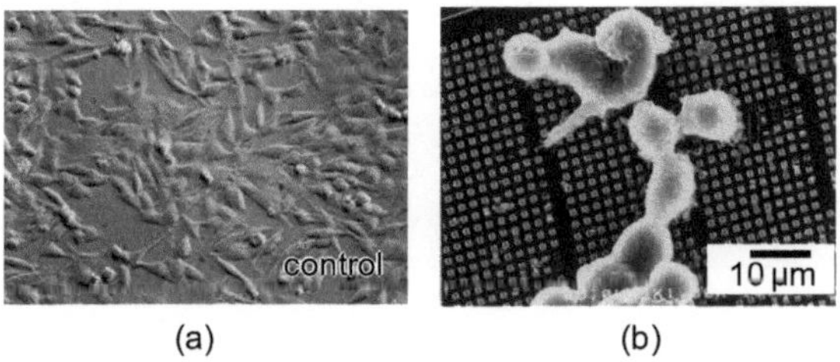

Figure 10.12 Grown cells on (a) conventional flat dish and (b) nanoimprinted scaffold.

Optical microscope images of HeLa cells grown on the nanopillars are shown in Fig. 10.13. The images in Figs. 10.13a and 10.13b show the cells before and after medium pipetting, respectively. The grown cells were easily removed by simply pipetting a medium. HeLa cells usually adhere to a culture dish strongly, so a proteinase such as trypsin (an enzyme that damages cells) is used to remove the attached grown cells from a conventional culture dish. Therefore, the difference in cell shape, shown in Fig. 10.12, is regarded as the difference in adhesion forces of the cells to the scaffold.

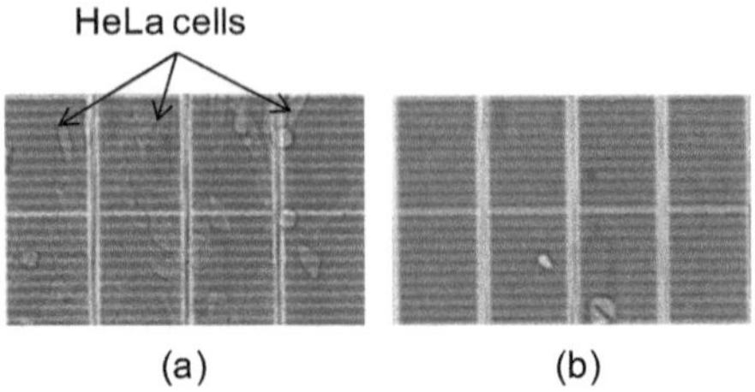

Figure 10.13 HeLa cell removal by pipetting a medium. Cells are removed easily by just pipetting a medium: (a) as grown and (b) after pipetting.

The growth characteristic on the nanopillar scaffold is also unique. Cartilage cell growths on flat and pillar surfaces are shown in Fig. 10.14. Cartilage cells are usually cultured by floating cultivation; therefore, as shown in Fig. 10.14a, chondrocytic cells cannot grow well on a flat scaffold. However, on the nanopillar scaffolds, cell clusters (spheroids) are formed. This spheroid formation is a unique characteristic of a nanopillar cell culture. A cell culture plate formed by nanoimprinting is described below.

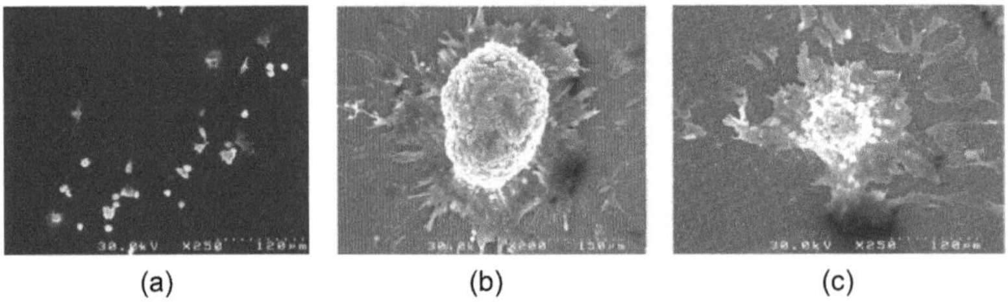

Figure 10.14 Growth characteristics of chondrocytic cells on various scaffold structures. (a) Conventional flat surface, (b) nanopillar with a diameter of 0.5 μm, (c) and nanopillar with a diameter of 2.0 μm.

The developed cell culture plate with 96 wells is shown in Fig. 10.15. The plate has 96 wells (in which cells are cultured), and 200 μm diameter holes—named "microincubators"—are formed on the bottom of each well. Furthermore, 2 μm diameter pillars are formed on the bottom of each microincubator. The cell culture plates are fabricated by thermal nanoimprinting. The template is silicon, and the substrate is a PS film. A silicon wafer is fabricated by photolithography. The 200 μm diameter microincubator holes and 2 μm diameter holes on the bottom of each microincubator were fabricated by two-step dry etching. A nickel replica mold was obtained from a silicon template by electroplating. The nickel replica mold was used to fabricate the microincubators by thermal nanoimprinting, as shown in Fig. 10.16. During the nanoimprinting, the holes and pillars are molded on the PS film in one go.

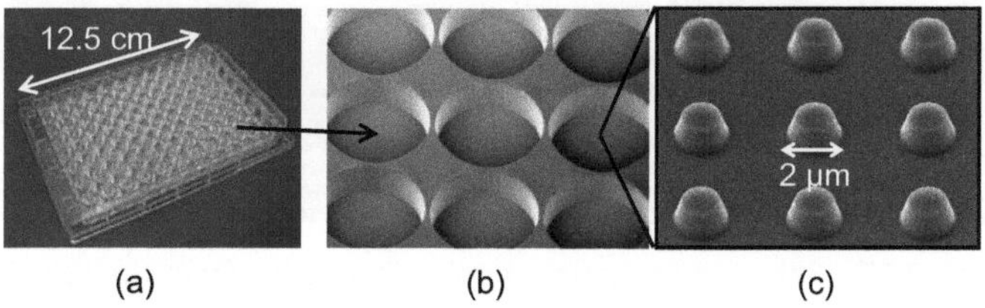

Figure 10.15 (a) Appearance of a nanopillar cell culture plate. (b) Microincubators with 200 µm diameter are fabricated on the well bottom. (c) Nanopillars with 2 µm diameter are fabricated on the bottom of the microincubator.

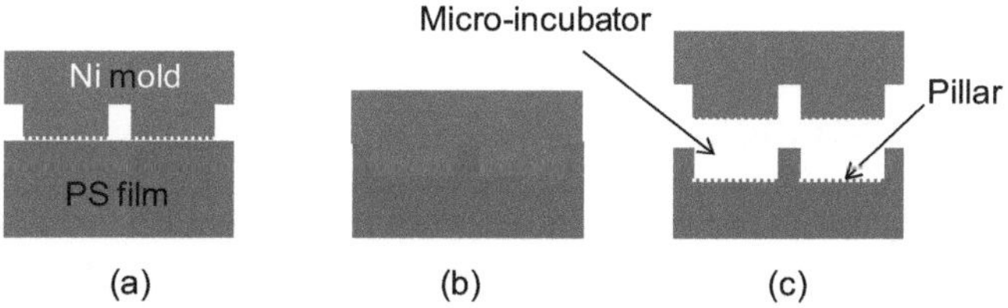

Figure 10.16 Process sequence of microincubator fabrication. (a) Prepare a Ni mold, (b) press the mold to a PS film, and (c) release the mold; microincubators are molded. PS, polystyrene.

Cultivation of hepatic cells is explained below. Hepatic cells are seeded in the wells of the culture plate and randomly sink into the microincubators. The cultivation states just after cell seeding and after 35 h and 70 h of cultivation are shown in Fig. 10.17 [15]. The cell is a hepatic cell of a rat. Almost the same numbers of cells were placed in the microincubators, and then spheroids of hepatic cells were generated. The pillar structures were formed in the regions indicated by the solid circular line. As cultivation time passes, the cells gather at the pillar regions. After cultivation for 70 h, spheroids were formed and most of them were located at the center of microincubator, as shown in Fig. 10.17c. The size of the spheroids is almost constant because the number of cells in each incubator was about the same. These spheroids function as metabolic reactions [16]. In other words, they act as a living liver. Therefore, for example, the effect of a newly developed medicine can be evaluated easily by using these hepatic spheroids.

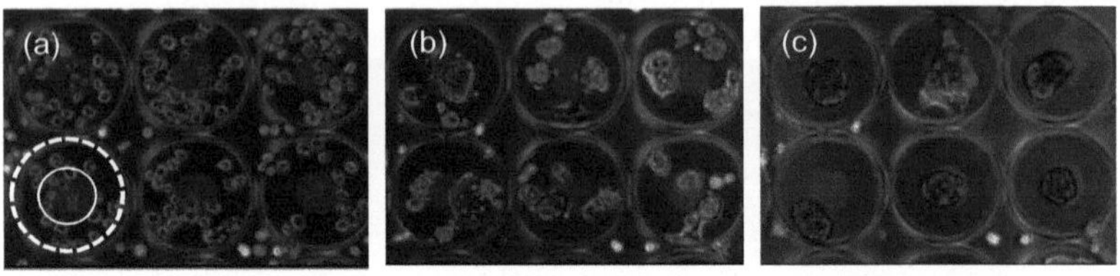

Figure 10.17 Spheroid formation in microincubators from the top view. The dashed line and solid line in (a) indicates the microincubator and pillar region, respectively. (a) Just after hepatic cell seeding, (b) after 35 h cultivation, and (c) after 70 h cultivation.

Three-dimensional cell cultivation is essential for regeneration medicine. It is the basis of tissue culture, and nanoimprinting will be one of the key technologies for structured scaffolds.

10.2.2 Immunoassay

The human body has an immune system to protect it from various infections, and an immunoassay is an important means of diagnosing such infections. The principle of a fluorescence immunoassay (FIA) is shown in Fig. 10.18. First, an antibody is bonded (solidified) to a substrate (step a). Then, a specimen is dropped onto the substrate. If an antigen is present in the specimen, an antigen-antibody reaction occurs (step b). An antigen-antibody reaction is a specific reaction. Next, a second antibody (which is labeled with a fluorescent marker) is dropped and the second antibody bonds to the antigen (step c). Finally, the fluorescence intensity is measured by irradiating excitation ultraviolet (UV) light (step d). By measuring the intensity of the fluorescence, it is possible to evaluate the density of the antigen in the specimen.

A nanoimprinted surface has a 3D structure, so its specific surface area is large. Therefore, a high fluorescence intensity is expected from a nanoimprinted surface. Accordingly, a protein was coated on a nanoimprinted surface, where the antigen-antibody reaction was triggered.

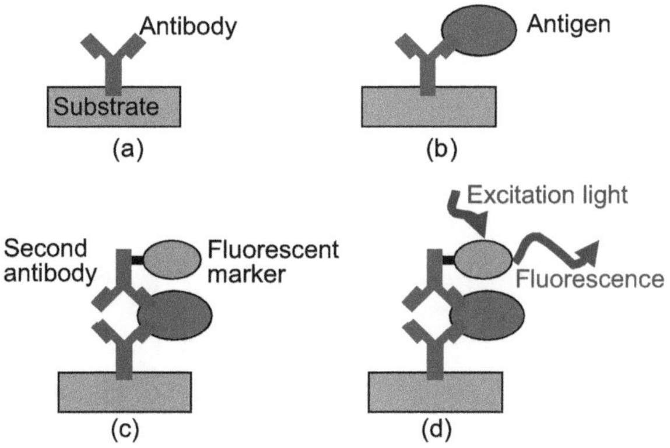

Figure 10.18 Principle of FIA. The diagnosis sequence is as follows: (a) the antibody coats the substrate, (b) the specimen is dropped and the antigens connect to the antibody, (c) a second antibody that has a fluorescent marker is dropped, and (d) excitation light is irradiated and fluorescence intensity detected. By measuring the fluorescence intensity, the antigen concentration can be obtained. FIA, fluorescence immunoassay.

A schematic diagram of a nanopillar surface coated with fluorescein-4-isothiocyanate (FITC)-albumin is shown in Fig. 10.19. FITC is a general fluorescence marker. The samples were prepared as follows: First, polymethylmethacrylate (PMMA) nanopillars were molded by thermal nanoimprinting. Then, the PMMA surface was irradiated by UV light to increase its wettability. Finally, the samples were immersed in an FITC-albumin solution and dried.

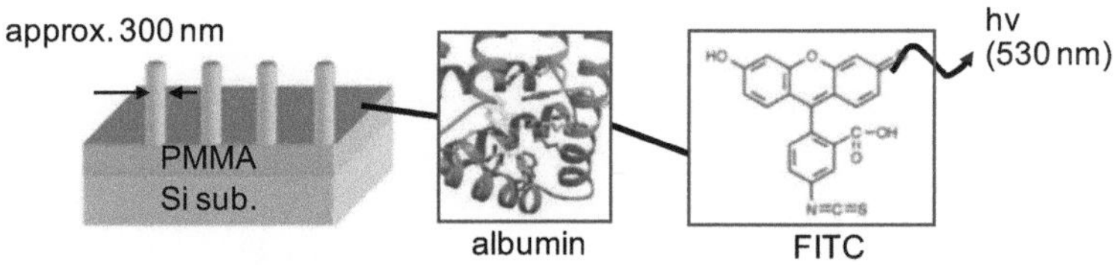

Figure 10.19 Schematic diagram of nanopillar surface covered with FITC-albumin. FITC, fluorescein-4-isothiocyanate.

A fluorescence image of the nanoimprinted surface is shown in Fig. 10.20. Various zones with specific surfaces are formed by various nanopillars on the chip, so bright and dark lines are observed on it (image a). In the magnified image (image b), the nanopillar region is brighter than the surrounding flat region. And each nanopillar shows fluorescence (image c). These images show that the nanopillar surface can be modified by albumin molecules. Also, other kinds of proteins are expected to modify the nanopillar surface. Therefore, an immunoassay reaction on nanopillars was investigated [17].

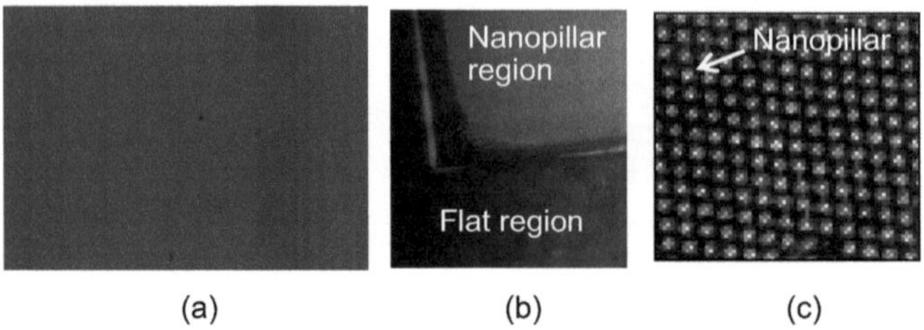

Figure 10.20 Fluorescence image from nanopillar surface. (a) Low-magnitude image, (b) near the boundary of the patterned region, and (c) high-magnitude image of nanopillar region.

Various kinds of nanopillar structures used in an immunoassay chip are shown in Fig. 10.21. Pillar diameters are 1 μm to 95 nm (the 1 μm nanopillar is square in shape). The aspect ratio reaches 43 in the case of structure d [18]. These high-aspect-ratio structures are obtained by elongation of the resin.

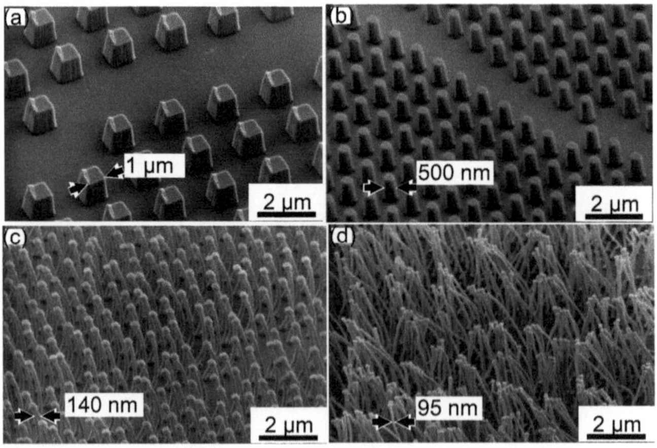

Figure 10.21 Various kinds of nanopillar structures for immunoassay chip.

A fluorescence image of a nanopillar immunoassay chip is shown in Fig. 10.22. Alpha fetoprotein (AFP), which is a marker of hepatic cancer, was used as the antigen. In Fig. 10.22, from right to left, the diameter of nanopillars decreases and the specific area increases. With an increasing specific area, the intensity of fluorescence increases. In other words, the reaction density of the antigen-antibody increases with the specific area, which is designed by nanoimprinting.

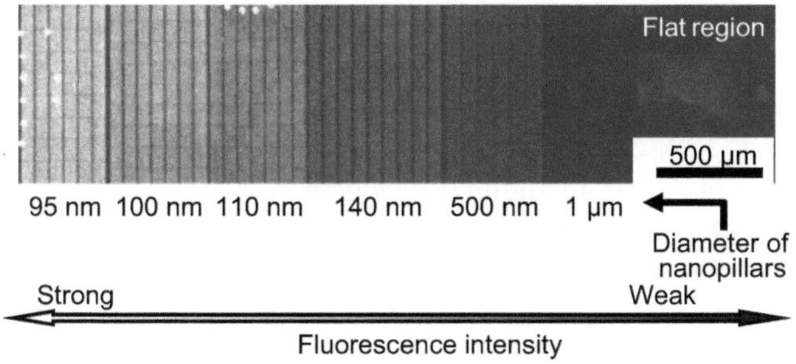

Figure 10.22 Fluorescence image from nanopillar surface. AFP concentration is 10^4 pg/mL. AFP, alpha fetoprotein.

Results of AFP measurements by a nanopillar immunoassay chip are plotted in Fig. 10.23. Clearly, the fluorescence intensity increases with increasing AFP concentration in each nanopillar region. The fluorescence intensity in the 95 nm nanopillar region is 34 times higher than that in the flat region.

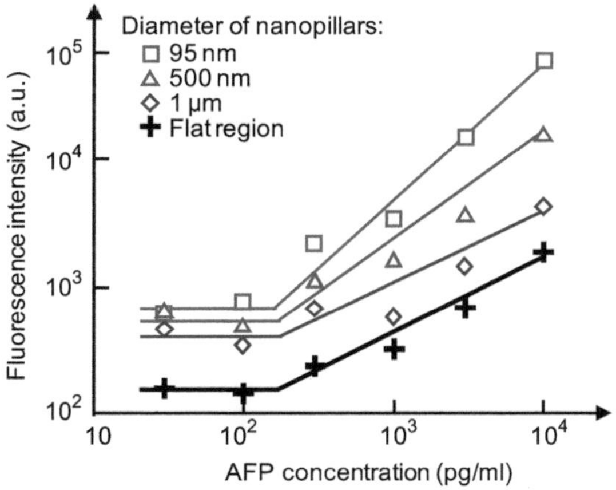

Figure 10.23 Fluorescence intensity as a function of AFP concentration in the specimen. AFP, alpha fetoprotein.

10.3 Applications to Energy Devices

In this section, two nanoimprint applications in energy devices, fuel cells and solar cells, are introduced.

10.3.1 Fuel Cells

Several types of fuel cells (FCs), such as a direct methanol fuel cells (DMFCs), polymer electrolyte fuel cells (PEFCs), phosphoric acid fuel cells (PAFCs), molten carbonate fuel cells (MCFCs), solid-oxide fuel cells (SOFCs), and alkaline fuel cells (AFCs), are well known. Any type of FC has an anode (fuel electrode), a cathode (air/oxidizer electrode), and an electrolyte between the electrodes. Ion carriers are generated at the surface of the fuel electrode, pass through the electrolyte, and react at the cathode electrode.

A schematic diagram of a DMFC is shown in Fig. 10.24. In the case of a DMFC, the supplied fuel (methanol) and water are decomposed on the cathode. Electrons flow via the external circuit to the anode electrode. Proton ions go through the electrolyte membrane and are combined with the electrons at the air electrode. Each chemical reaction is generated on both surfaces of the electrolyte membrane. As explained in Section 10.2.2, the chemical reaction density is enhanced by the increasing specific surface area. Therefore, power generation per unit volume is expected to increase on using a nanoimprinted electrolyte membrane.

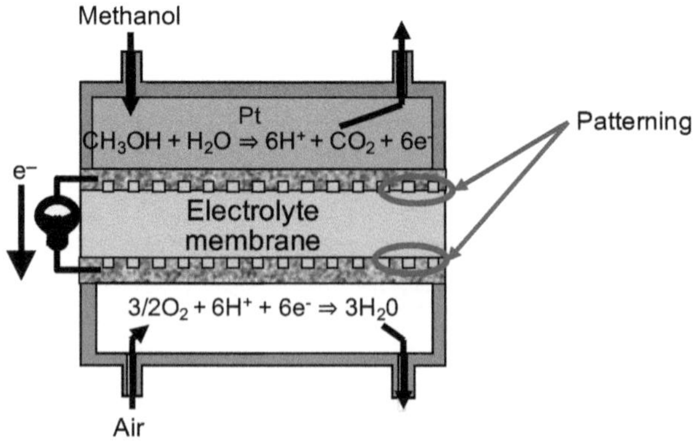

Figure 10.24 Schematic diagram of nanoimprinted DMFC. The surfaces of the electrolyte membrane are patterned by thermal nanoimprinting. DMFC, direct methanol fuel cell.

The interface between a fabricated electrolyte membrane (made of Nafion®) and a catalytic layer is shown in Fig. 10.25. The pillars on the membrane surface were molded by thermal nanoimprinting. The catalytic layer was deposited by a spray method, so there must be a specific amount of space between the pillars to be filled with catalytic particles.

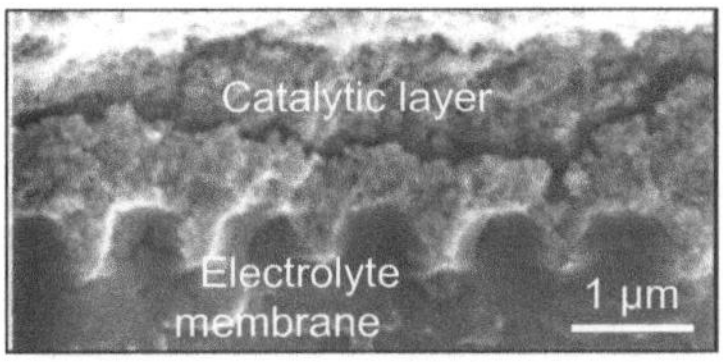

Figure 10.25 Cross-sectional SEM image near interface between patterned electrolyte membrane and catalytic layer. SEM, scanning electron microscopy.

Power generation characteristics of a DMFC are plotted in Fig. 10.26. When an imprinted membrane is used, the power generation density becomes about two times higher than that of a conventional plane membrane. This increase is due to an increased surface-reaction density and reduced contact resistance between the membrane and the catalytic layer.

The DMFC described in the figure was developed as part of the New Energy and Industrial Technology Development Organization (NEDO), Japan.

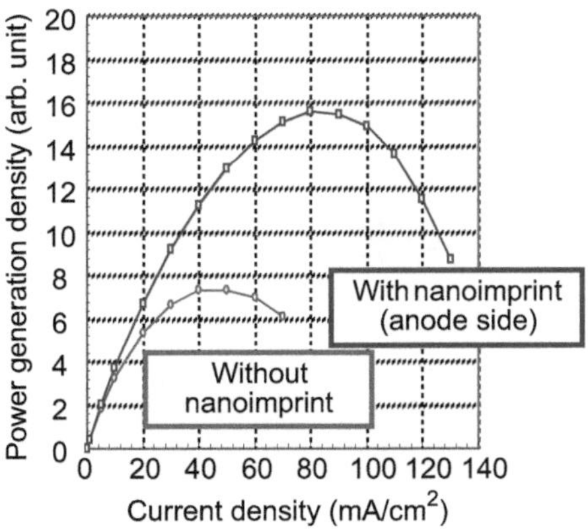

Figure 10.26 Power generation characteristics of nanoimprinted DMFC. DMFC, direct methanol fuel cell.

10.3.2 Solar Cells

As described in Section 10.1, nanoimprinting is suitable for fabricating optical devices. In view of energy devices, a solar cell is also an application field of nanoimprinting because controlling the light path is important for attaining high cell efficiency. Controlling the light path in a solar cell is introduced next.

A cross-sectional diagram of a typical solar cell is shown in Fig. 10.27. Sunlight incident on the top surface is absorbed on the semiconductor layers. Electron-hole pairs are generated in the semiconductor layers by the light absorption. Electrons and holes are separated in the depletion layer at the interface of the p- and n-type semiconductor layers. A difference in the Fermi levels of the p- and n-type semiconductors causes an electromotive force. The thickness of the semiconductor layers is several microns, so light absorption in the semiconductor layer becomes important.

Confinement states of sunlight in the semiconductor layer are shown in Fig. 10.28. After the sunlight enters the top electrode, it is reflected at the surface of the back electrode. The behavior of the reflected light depends on the scattering status at the back electrode. When the back electrode is a flat surface, the reflected light transmits back to the top electrode, as shown in Fig. 10.28a. However, if the surface of the back electrode is rough, the incident light is scattered the way it is shown in Fig. 10.28b. Furthermore, if the incident angle of the reflected light to the transparent electrode is greater than the total reflection angel, θ_m, the reflected light will be confined to the thin semiconductor layer. Accordingly, the light scattering status was measured using a nanoimprinted substrate.

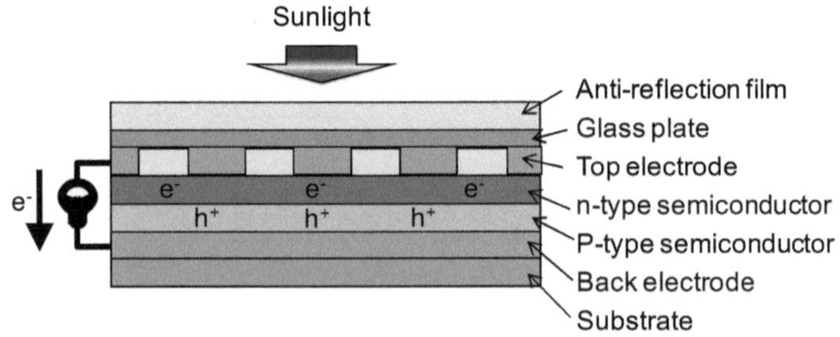

Figure 10.27 Schematic diagram of a solar cell. p- and n-type semiconductor layers are sandwiched by electrodes.

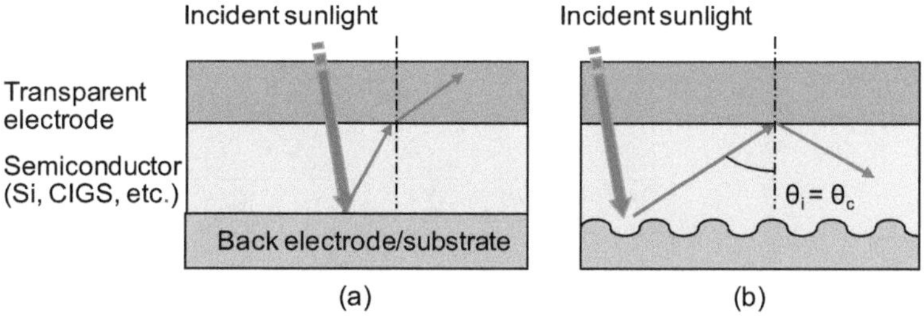

Figure 10.28 Confinement states of sunlight in semiconductor layer. (a) Plane surface and (b) textured surface.

The morphology of the nanoimprinted substrate is shown in Fig. 10.29. It is made of a polyimide, which has high heat durability, so it can withstand the silicon deposition process. The polyimide was patterned by thermal nanoimprinting using an imide oligomer as the source material.

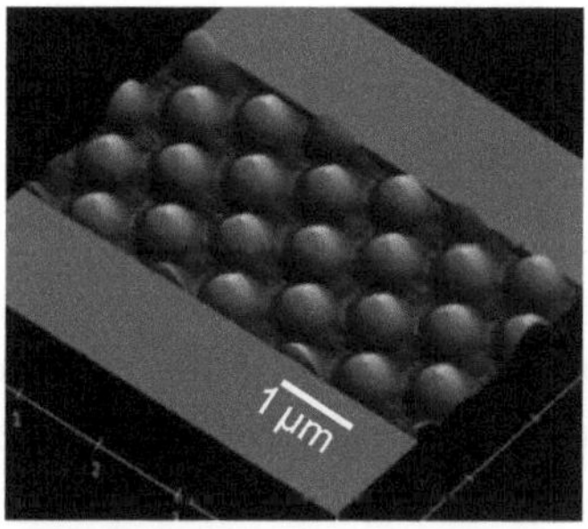

Figure 10.29 Morphology of patterned polyimide surface observed by AFM. The measured area is limited, but the convex patterns are molded in the whole substrate area. AFM, atomic force microscopy.

The measured brightness of the scattering light is plotted as a function of the scattering angle in Fig. 10.30. The incident light irradiated the back electrode (substrate) at an incident angle of 0°. The light is scattered at an angle of more than about 35°. The total reflection angle is 28° in the case of an indium tin oxide (ITO) and the silicon interface. Therefore, by using this patterned substrate, it is possible to confine the reflected light from the back electrode to the thin semiconductor layer.

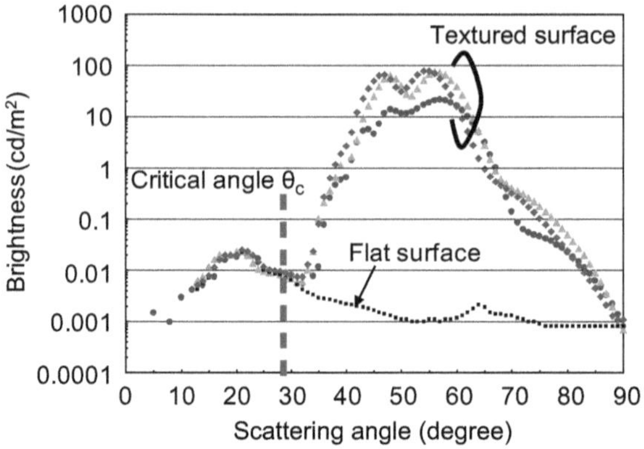

Figure 10.30 Brightness of scattering light as a function of scattering angle. Total reflection occurs when the incident angle becomes more than 28° in the case of the ITO/Si interface. ITO, indium tin oxide.

10.4 Electric Devices

Large-scale integration (LSI) circuits are conventionally produced by photolithography. However, instead of photolithography, photonanoimprinting is expected to be applied to the fabrication [19–21]. Nanoimprinting machines for LSI circuit production have already been commercialized by Canon Nanotechnologies, Inc., and related technologies are being improved.

In this section, the application of photonanoimprinting in fabricating a future device, a "memristor," is introduced [22, 23]. A memristor is a fourth-class passive device. It has a metal/insulator/metal (MIM) structure, in which the insulator acts as a switching material. Photonanoimprinting is applied to fabricating the crossbar-array structure of the memristor.

A schematic diagram of the fabrication sequence of a memristor is shown in Fig. 10.31. A quartz mold, composed of nanowires with a 100 nm half pitch and microscale fan-outs, is prepared by electron beam lithography (EBL) and photolithography, respectively. A double layer of resin is spin-coated on the substrate (step a). The substrate is a silicon wafer with a 100-nm-thick SiO_2 layer. The top resist is patterned by photonanoimprinting using the prepared quartz mold (step b). Then, a 3.5-nm-thick titanium/9-nm-thick platinum layer is deposited by shadow evaporation (step c). This layer becomes the bottom metal layer of the MIM structure. Next, a 13-nm-thick TiO_2 layer is deposited by DC sputtering (step d). Then, a 12.5-nm-thick platinum electrode is deposited by shadow EB evaporation. Finally, the top resist is removed.

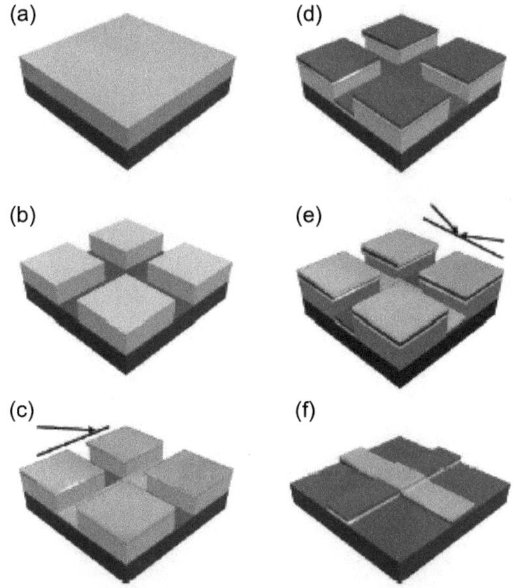

Figure 10.31 Schematic of the self-aligned fabrication approach with one NIL step. The cross-shaped trenches are patterned by one NIL step and RIE in resists; after depositing the three layers, a liftoff process concludes the fabrication. The metal electrodes were deposited using angle evaporation so that the metals reach the bottom of one trench but not the other. (a) Spin-double-layer resists on the substrate, (b) NIL with a cross-bar mold and RIE, (c) deposit bottom electrodes using shadow evaporation, (d) deposit-switching materials layer using sputtering, (e) deposit top electrodes using shadow evaporation, and (f) liftoff in solvent. The arrows in (c) and (e) indicate the shadow evaporation directions. NIL, nanoimprint lithography; RIE, reactive ion etching. Reprinted with permission from Ref. [22]. Copyright (2010) American Chemical Society.

Images of a 1 × 21 array of memristors are shown in Fig. 10.32. The layout of nanowires is manufactured in the mold; therefore, no alignment process is required in the memristor fabrication process.

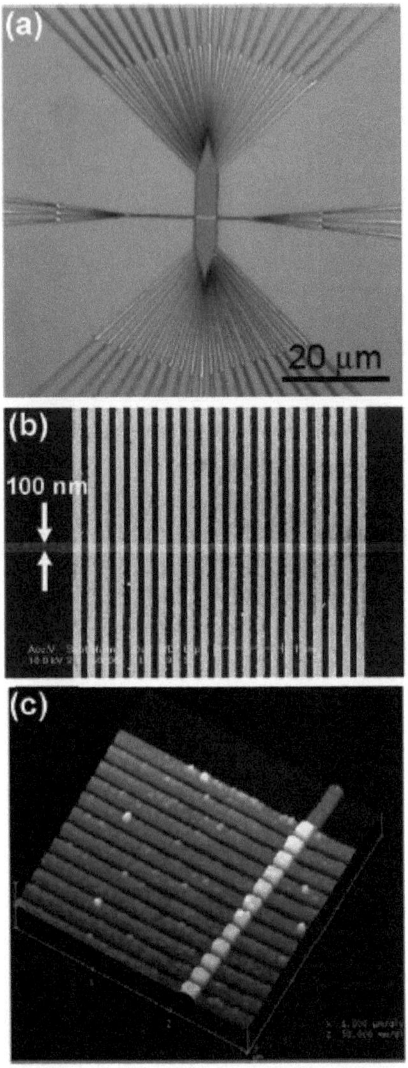

Figure 10.32 Images of a 1 × 21 array of memristors fabricated using one NIL step. (a) Optical microscopy image. The three fan-out microwires in the horizontal direction are connected to a single nanowire. (b) SEM image of the cross-point array. The junction area for each device is 100 × 100 nm^2. (c) AFM image of part of the array. NIL, nanoimprint lithography; SEM, scanning electron microscopy; AFM, atomic force microscopy. Reprinted with permission from Ref. [22]. Copyright (2010) American Chemical Society.

The typical switching behavior of fabricated memristors is shown in Fig. 10.33. The memristors exhibit a nonvolatile switching behavior with an on/off ratio greater than 1000 at 1 V.

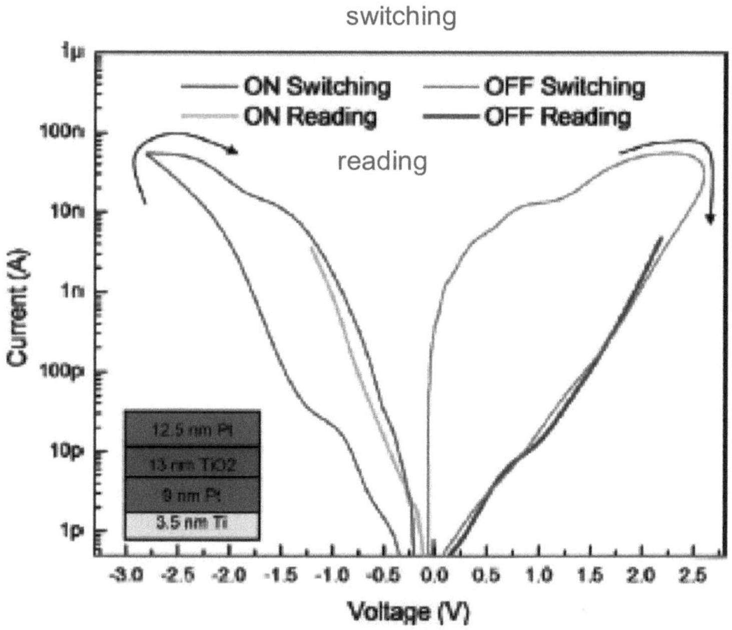

Figure 10.33 Typical switching behavior of memristive devices fabricated using one NIL step with oblique deposition angles of 20°. The devices were fabricated on a Si wafer with 100-nm-thick thermal oxide with 13-nm-thick TiO_2 sandwiched between the 12.5-nm-thick Pt top electrodes and the 9-nm-thick Pt/3.5-nm-thick Ti bottom electrodes (geometry shown in the inset; not to scale). The device junction area is 100 × 100 nm^2. The devices exhibit nonvolatile switching behavior with an ON/OFF ratio larger than 1000 at 1 V. No forming step was necessary, and the switching current was much lower than usually observed for similar devices fabricated using a multistep lithography process. NIL, nanoimprint lithography. Reprinted with permission from Ref. [22]. Copyright (2010) American Chemical Society.

In this field, 8 × 8 nm memristive devices have been fabricated by photonanoimprinting [23]. Nanoimprint applications in the electric device field will expand and make it possible to fabricate the next generation of electronic devices.

10.5 Nanoimprint Lithography for the High-Volume Manufacturing of Advanced Semiconductor Devices

10.5.1 Introduction

Optical lithography has been the technology of choice for many decades for the manufacturing of advanced large-scale integration (LSI) semiconductor devices. The most advanced tools currently used are immersion-based ArF scanner systems capable of resolving features as small as 40 nm. However, today's most advanced devices require half pitches as small as 14 nm, and complex and costly pitch reduction methods, including self-aligned multiple patterning, must be applied to meet these requirements. A lithographic technique that can directly pattern these features less than 20 nm is highly desired.

Extreme ultraviolet (EUV) lithography is being developed but will require extremely high-cost tools with limited throughput and high running costs. Therefore, the cost advantage of such tooling is limited.

Nanoimprint lithography (NIL) is also being developed for this application and provides direct patterning at a low cost, with reduced running costs. NIL is a high-throughput, high-resolution parallel patterning method in which the relief images in a template (mask or stamp) are replicated into a material by mechanical contact and material displacement [24–26]. This can be done by shaping a liquid followed by a curing process for hardening. The most common NIL process requires heat input in order to allow the imprint resist to flow into the relief images of a mask.

Devices that require several lithography steps and precise overlay will need an imprinting process capable of addressing registration issues. A derivative of NIL, ultraviolet nanoimprint lithography (UV-NIL) solves the issue of alignment by using a transparent template, thereby facilitating conventional overlay techniques. In addition, the imprint process is performed at low pressure and at room temperature, which minimizes magnification and distortion errors. Canon utilizes an NIL process that locally addresses issues of residual layer thickness nonuniformity by locally dispensing a low-viscosity resist material. This second approach was first disclosed by Willson et al. in 1999 [27].

The process involves the field-by-field deposition and exposure of a low-viscosity resist deposited by a drop-on-demand inkjet onto the substrate [28–36]. The patterned mask is lowered into the fluid, which then quickly flows into the relief patterns in the mask by capillary action. Following this filling step, the resist is cross-linked under UV radiation and the mask is removed, leaving a patterned resist on the substrate.

To successfully apply NIL for semiconductor device manufacturing, new apparatus is required to address throughput, alignment, overlay, defectivity, and mask replication. For example, NIL does not employ a lens system capable of correcting in-field distortions. Therefore, a different approach is needed to address high-order distortion correction (HODC). The purpose of this review paper is to report the progress made in addressing the key parameters noted above. In Section 10.5.2, wafer imprint attributes are discussed. In Section 10.5.3, mask replication tooling performance is addressed. Following these sections are some brief conclusions and acknowledgment.

10.5.2 Wafer Imprint Tool

The configuration of the FPA-1200 NZ2C wafer imprint tool is shown schematically in Fig. 10.34. A four-station cluster configuration is used to enable high throughput. Specifications for the current system are 80 wafers per hour (wph) (meaning 20 wph per station), with an overlay specification of 4 nm, 3 sigma and a particle adder specification of 0.0008 particle adders per wafer pass (or approximately 1 particle every 1200 wafers). Throughput, alignment and overlay, and defectivity and particle adders are discussed in the following sections.

10.5.2.1 Wafer throughput

The standard NIL sequence for both a 60 wph and an 80 wph process is shown in Figure 10.35. After resist dispense, the mask is brought into contact with the liquid resist. Alignment and resist filling are done in parallel. Following this step, the resist is exposed and the mask is separated from the resist. Clearly, the limiting step in the process is resist filling. For example, for an 80 wph process, the resist fill step consumes 40% of the throughput budget.

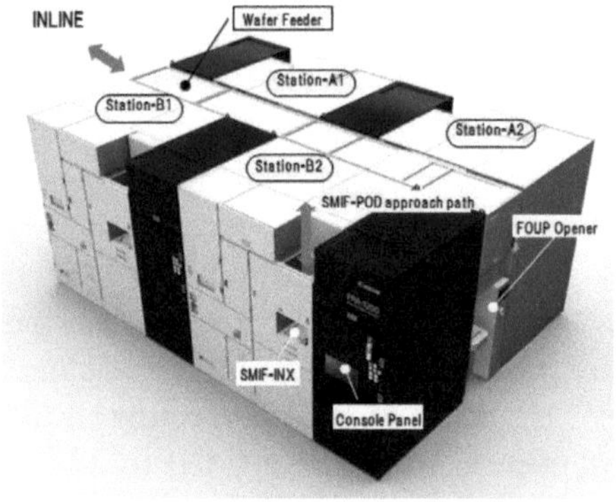

(a)

Current performance	
Throughput	80wph
Mix-and-Match Overlay	4.0nm
Particle Adder	0.0008 pcs/wafer

(b)

Figure 10.34 FPA-1200 NZ2C cluster NIL system for mass production. Current specifications are 80 wafers per hour, with an overlay specification of 4 nm, 3 sigma and a particle adder specification of 0.0008. NIL, nanoimprint lithography.

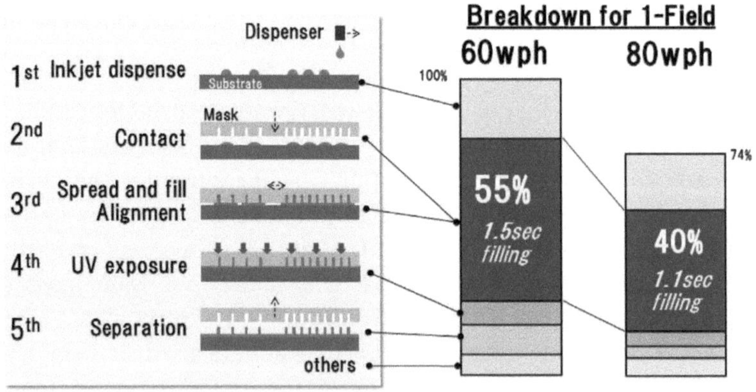

Figure 10.35 NIL throughput breakdown. The critical parameter in meeting throughput specifications is resist filling. NIL, nanoimprint lithography.

There are several parameters that can impact resist filling. Key parameters are:

- Resist drop volume: Smaller is better (<1 pL) since smaller volumes allow resist droplets to be placed closer together.

- Material engineering (to promote wetting between the resist and the underlying adhesion layer).
- Design for imprint (DFI; to accelerate drop spreading and address different pattern types).
- System controls that address drop spreading after jetting for both full fields and partial fields.
- Drop pattern optimization.

In addition, it is mandatory to maintain fast filling, even for edge field imprinting. Previously, we have demonstrated that it is feasible to fill dense line-and-space (L&S) patterns in only 1 s.

The resist properties have a large impact on fill time, and engineering of the resist is critical for meeting performance criteria and properties such as surface tension, viscosity, and wetting. Surface wetting has a strong influence on fill time and has been addressed with newer resist formulations that cause resist spreading to increase by a factor of 3 before the mask makes contact with the resist material on the wafer surface [37].

As a rule of thumb, a 1.1 s fill time is necessary to achieve a throughput of 20 wph per imprint station or 80 wph for a fully configured four-station NZ2C cluster tool. In 2016, throughputs of 15 wph with a 1.5 s fill time were achieved. More recently, 1.1 s filling on device-like layouts has been achieved.

10.5.2.2 Alignment and overlay

The alignment and overlay system consists of various factors, which can be categorized generally as alignment and distortion. In a multistation system, such as the NZ2C, it is important to recognize that overlay and distortion must be controlled within a station and from station to station. Additionally, the magnification actuator system initially employed for overlay is limited in its ability to correct for high-order distortion signatures present on the previous pattern levels of a device. Thus a method for correcting these distortions is required.

Using a through-the-mask (TTM) Moiré-based die-by-die alignment system [38], 1 nm repeatability has been demonstrated and the data collected by the TTM system correlate very closely with an Archer measurement tool. A schematic diagram of Canon's TTM system is shown in Fig. 10.36. The two basic components of the TTM alignment system are relay optics and a TTM scope. The relative positions of the mask mark and the substrate mark are measured by measuring the reflected Moiré signal in the TTM scope.

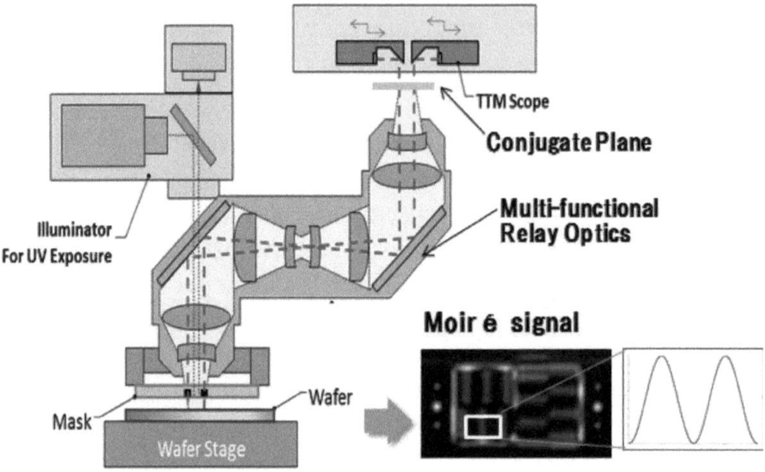

Figure 10.36 Schematic drawing of the TTM alignment system. The TTM alignment system comprises relay optics and TTM scope. A typical Moiré pattern is shown to the right. TTM, through-the-mask.

Canon has studied both single-machine overlay (SMO) and mix-and-match overlay (MMO). SMO measurements are not done in the same way as an optical projection system, since it is not possible to print a second time over the existing resist. As a result, the wafer must be removed after a first imprint and etched and placed back in the tool for the subsequent imprinting. Across the wafer, SMO was less than 2.2 nm. The best MMO data are shown in Fig. 10.37. Across the wafer, 4 nm, 3 sigma were observed.

To minimize overlay errors, it is necessary to address both magnification and distortion corrections. The basic system used by Canon consists of two components, as shown in Fig. 10.38. As shown in the top half of the figure, magnification, skew, and trap errors can be addressed by applying a magnification actuator system consisting of an array of piezo actuators that push against the sides of the mask.

In optical lithography, high-order distortions can be corrected through manipulation of the projection lens and the wafer scan stage. However, in an NIL apparatus, there is neither a projection lens nor a scan wafer stage. Therefore, an HODC system was developed to correct distortion errors on the wafer. The HODC process is shown in the bottom half of Fig. 10.38.

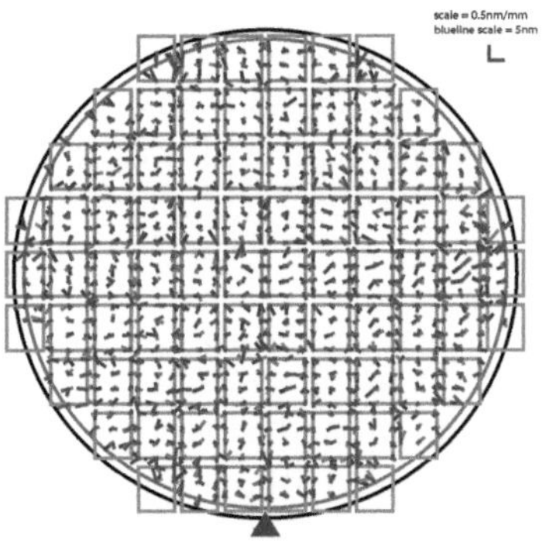

Tool : FPA-1200 NZ2C
Mask : Test mask with 1x nm features
Substrate : ArF IML
n of fields : 84 (full and partial)
n of measured points : 12 / field

3wafers total [nm]

	X	Y
|m|	0.1	0.0
3σ	3.9	4.0
|m|+3σ	**4.0**	**4.0**

Figure 10.37 Mix-and-match overlay to a 193 nm immersion scanner.

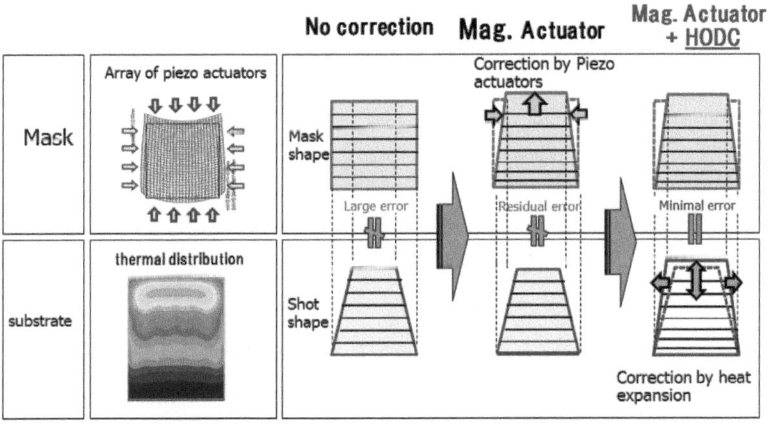

Figure 10.38 Schematic illustration of the magnification and distortion correction system used in the FPA-1200 NZ2C nanoimprinting system. A magnification actuator system corrects magnification errors. A die-by-die HODC system is used to locally heat each field on the wafer to correct higher-order distortion terms. HODC, high-order distortion correction.

The system functions by applying a heat input on a field-by-field basis using a digital micromirror device (DMD) and defines a potential path for achieving overlay results better than sub 3.5 nm. Initial heat input experiments have now been performed, and the results are very promising. Simulation and experimental results are in excellent agreement, and reproducibility across a four-wafer set of ~1 nm was demonstrated. Testing was first done on the NZ2C tool, first looking at the ability to correct known high-order terms such as K7 (2nd Mag), K11 (Bow), and K17 (Pincushion). More recent work where HODC was applied to adjust for K7, K8, K11, K12, K13, K14, K17, and K18 distortions resulted in residual errors of less than 1 nm for every case.

10.5.2.3 Defectivity and mask life

Figure 10.39 shows a cost comparison between conventional optical lithography and NIL. As previously mentioned, direct printing of critical features reduces cost as compared to lithographic methods requiring multiple patterning techniques, such as self-aligned double patterning (SADP) and self-aligned quadruple patterning (SAQP). Each bar in the figure contains three different components: (i) the tool cost, shown in blue, (ii) the mask cost, in red, and (iii) the additional process cost, in green.

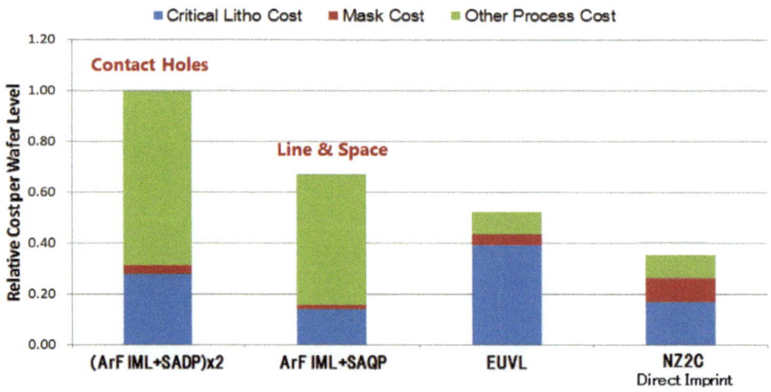

Figure 10.39 Relative cost per wafer level. Direct nanoimprint lithography offers the lowest cost of ownership.

NIL offers the lowest cost of ownership (CoO); however, it is important to note that mask cost plays an important role in NIL CoO reduction. The figure assumes that a replica mask (more details about the replica mask are described in the next section) will last for 40 lots, which is equivalent to 1000 wafers. Mask life, in turn, depends on the interaction between the mask and either the mask or the wafer. In general, soft particles do not cause permanent mask damage. Hard particles can induce mask damage, thereby causing a permanent increase in mask defectivity. Therefore, it is critical to reduce the number of particles introduced during the imprint process.

Within the NIL tool there are two primary particle sources. The first is particles originating from the resist delivery system. These are easily addressed by employing a recirculating resist system.

The more challenging task is to minimize particles introduced within the imprint tool itself. To meet the CoO specifications required for memory devices, the replica mask life must be sustained for better than 1000 wafers. If we conservatively assume that

- Every hard particle adds a defect to the mask and
- The mask defectivity limit from hard particles is 0.1 pieces per square centimeter, then the number of particle adders per wafer pass must be <0.001. As a result, if we are to achieve this particle specification, an aggressive strategy is needed to remove particle adders to the wafer and the mask. The methodology for removing particles can be broken down into categories:
- Particle source reduction: This means the minimization of particle generation from particle sources related to materials within the tool and the surface treatment of these materials. For example, ceramics are a known source of particles and proper treatment of the surface of ceramic parts can greatly reduce particle generation.
- Removal: The reduction of particles that could potentially find their way onto the mask and wafer. These can be addressed by optimizing the airflow within the tool. Figure 10.40 shows a schematic illustration of an air curtain that has been applied to the tool. The air curtain works to direct any existing particles in the system away from the mask and wafer.
- Particle mitigation: For example, electrostatic plates have been inserted to direct any particles that do end up near the mask and wafer to an area adjacent to the mask and wafer.

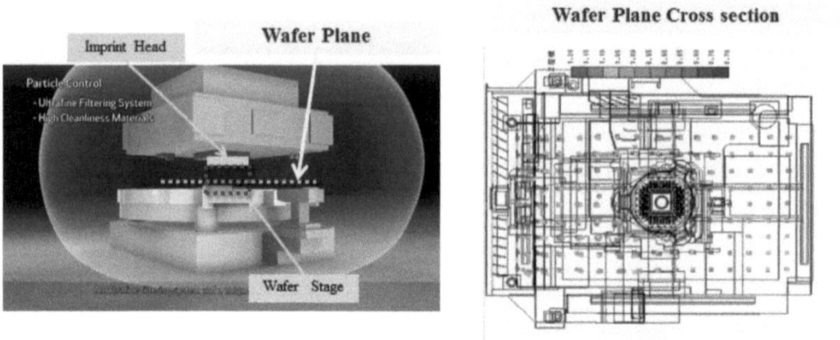

(a) (b)

Figure 10.40 (a) Schematic diagram of the imprint tool and (b) a simulation of particles tracked within the imprint tool.

The implementation of these particle reduction methods has enabled a reduction in particle adders of more than 4 orders of magnitude in four years, exceeding the targeted tool adder specification of 0.001, as shown in Fig. 10.41.

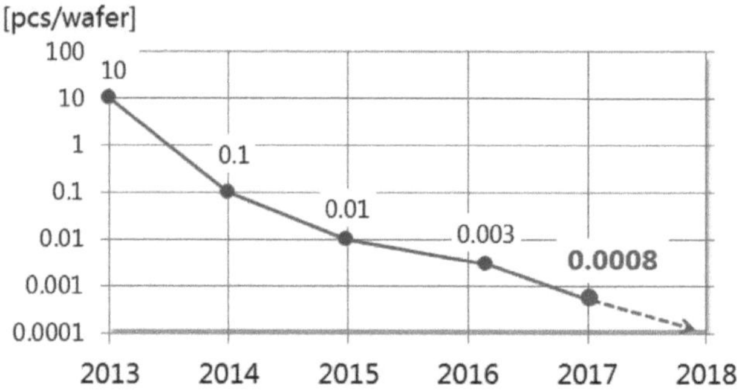

Figure 10.41 Particle adder reduction over the course of four years.

10.5.3 Mask Replication Systems

Because mask lifetime is limited in NIL, a low-cost method is required for manufacturing replica masks made from an original master mask patterned using electron beam lithography. Canon's FPA-1100 NR2 is designed to specifically address this task. The tool is shown schematically in Fig. 10.42. Tool throughput is 4 masks per hour, critical dimension (CD) uniformity is 0.8 nm, and image placement (IP) accuracy is 1 nm. CD uniformity and IP are discussed in the sections that follow.

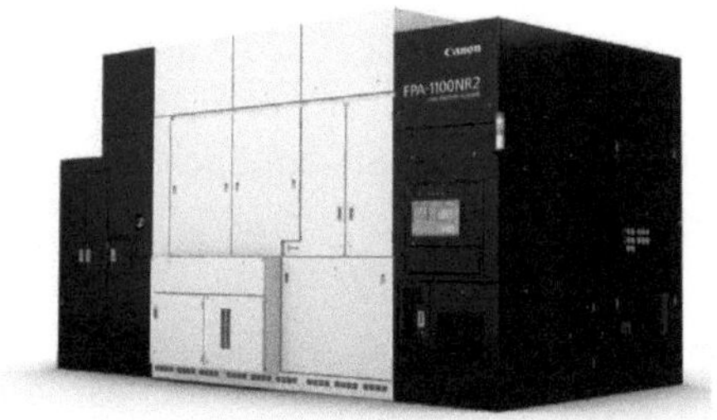

Current performance	
Throughput	4Mask/hour
CD Uniformity (3Ã)	0.8nm
Image Placement Accuracy (3Ã)	1.0nm
Particle	0.002pcs/Mask

Figure 10.42 The FPA-1100 NR2 mask replication system and specifications.

10.5.3.1 Critical dimension uniformity

The mask replication process requires good control of the critical features, IP accuracy, and defectivity. Critical feature control and defectivity are the subject of previous papers [39, 40]. Examples of imprinted 19 nm half-pitch features are shown in Fig. 10.43. Pictured are 19 nm L&S pattern SEM images at five positions in a 26 mm × 33 mm imprint field. CD uniformity is 1.3 nm, 3 sigma.

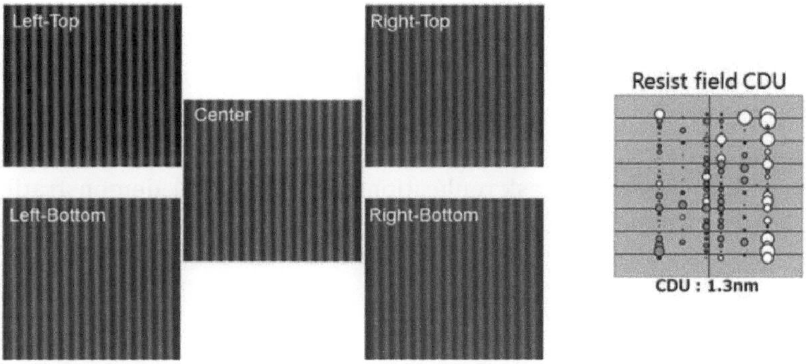

Figure 10.43 Examples of imprinted 19 nm half-pitch features. Pictured are 19 nm line-and-space pattern SEM images at five positions in a 26 mm × 33 mm imprint field. Critical dimension uniformity is 1.3 nm, 3 sigma. SEM, scanning electron microscopy.

10.5.3.2 Image placement

Equally important is the IP error induced during the replication process. Current IP specifications for sub 15 nm devices call for an added IP error on the replica to be less than 1 nm. To optimize IP accuracy, we improved chuck flatness and tilt control of the master and replica plate and also optimized the imprint sequence.

10.5.4 Conclusions

Great progress has been made in the field of NIL over the last four years. In this paper, to meet CoO requirements and to address yield issues, the progress on particle reduction, throughput, and overlay was reported. The continued reduction of particle adders extends the life of both the master mask and the replica mask. In this work, an air curtain system was tested both on a test stand and on the imprint tool. Optimization of this new scheme as well as enhancements to the existing particle control systems will aid in further reduction of imprint defectivity in order to meet the levels required for device manufacturing.

Throughputs of up to 20 wph per imprint station have also been achieved. This translates to 80 wph for a four-station cluster tool and is considered sufficient for high-volume manufacturing.

An MMO of 4 nm has been demonstrated, and methods for reducing overlay error using heat input through an HODC system were introduced. The HODC system has successfully corrected high-order terms, up to K20.

Finally, data on mask replication were presented, demonstrating an ability to minimize CD and IP errors.

Appendix D: Bragg's Law

A conventional antireflection coating is obtained by using Bragg's law. The reflection coefficient, R, is generally expressed as

$$R = 1 - T, \tag{10.1}$$

where T is the transmission coefficient.

The incident light is reflected at the interface between materials with different refractive indices. Accordingly, as shown in Fig. 10.44, the incident light is reflected at the surface of material 1 and the interface between material 1 and material 2.

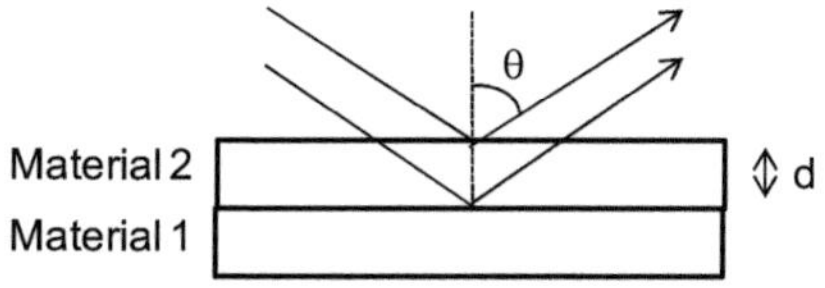

Figure 10.44 Schematic diagram of Bragg reflection.

The differing optical paths of the two lights are given as $2d\sin\theta$, where θ is the incident angle of the light. When $2d\sin\theta$ becomes $n\lambda$ (λ is the wavelength of the light), the reflected lights are intensified together. This state is called the Bragg condition, expressed as

$$2d\sin\theta = n\lambda. \tag{10.2}$$

If the optical path difference is $(n + \lambda/2)$, the reflected lights cancel each other out. An antireflection coating can thus be created by exploiting this light-canceling phenomenon.

References

1. A. Yokoo, H. Suzuki, and M. Notomi (2004). Organic photonic crystal band edge laser fabrication by direct nanoprinting, *Jpn. J. Appl. Phys.*, **43**(43), 4009–4011.
2. K. Watanabe, T. Morita, R. Kometani, T. Hoshino, K. Kondo, K. Kanda, Y. Haruyama, T. Kaito, J. Fujita, M. Ishida, Y. Ochiai, T. Tajima, and S. Matsui (2004). Nanoimprint using three-dimensional microlens mold made by focused-ion-beam chemical vapor deposition, *J. Vac. Sci. Technol. B*, **22**(1), 22–26.
3. H. K. Cho, J. Jang, J.-H. Choi, J. Choi, J. Kim, J. S. Lee, B. Lee, Y. H. Choe, K.-D. Lee, S. H. Kim, K. Lee, S.-K. Kim, and Y.-H. Lee (2006). Light extraction enhancement from nanoimprinted photonic crystal GaN-based blue light emitting diodes, *Opt. Express*, **14**, 8654–8660.
4. K. Sogo, M. Nakajima, and Y. Hirai (2006). Fabrication of anti-reflective structure by nano casting method, *J. Photopolym. Sci. Technol.*, **19**(5), 647–651.
5. W. Wu, M. Hu, F. S. Ou, Z. Li, and R. S. Williams (2010). Cones fabricated by 3D nanoimprint lithography for highly sensitive surface enhanced Raman spectroscopy, *Nanotechnology*, **21**, 255502.

6. Y.-K. R. Wu, A. E. Hollowell, C. Zhang, and L. J. Guo (2013). Angle-insensitive structural colours based on metallic nanocavities and coloured pixels beyond the diffraction limit, *Sci. Rep.*, **3**, 1194.
7. C. Zhang , S.-L. Chen, T. Ling, and L. J. Guo (2015). Review of imprinted polymer microrings as ultrasound detectors: design, fabrication, and characterization, *IEEE Sens. J.*, **15**, 3241–3248.
8. J. J. Wang, L. Chen, X. Liu, P. Sciortino, F. Liu, F. Walters, and X. Deng (2006). 30-nm-wide aluminum nanowire grid for ultrahigh contrast and transmittance polarizers made by UV-nanoimprint lithography, *Appl. Phys. Lett.*, **89**, 141105.
9. Y. Zhaoning, P. Deshpande, W. Wu, J. Wang, and S. Y. Chou (2000). Reflective polarizer based on a stacked double-layer subwavelength metal grating structure fabricated using nanoimprint lithography, *Appl. Phys. Lett.*, **77**, 927–929.
10. T. Senn, O. Kutz, and C. Weniger (2011). Integration of moth-eye structures into a poly(dimethylsiloxane) stamp for the replication of functionalized microlenses using UV-nanoimprint lithography, *J. Vac. Sci. Technol.*, **B29**, 061601.
11. H. Cao, Z. Yu, J. Wang, J. O. Tegenfeldt, R. H. Austin, E. Chen, W. Wu, and S. Y. Chou (2002). Fabrication of 10 nm enclosed nanofluidic channels, *Appl. Phys. Lett.*, **81**, 174–176.
12. L. Jay Guo, X. Cheng, and C.-F. Chou (2004). Fabrication of size-controllable nanofluidic channels by nanoimprinting and its application for DNA stretching, *Nano Lett.*, **4**, 69–73.
13. S. Nomura, H. Kojima, Y. Ohyabu, T. Uemura, K. Kuwabara, and A. Miyauchi (2005). Cell culture on nanopillar sheet: study of HeLa cells on nanopillar sheet, *Jpn. J. Appl. Phys. Part 2*, **44**(37–41), L1184–L1186.
14. S. Nomura, K. Kojima, Y. Ohyabu, K. Kuwabara, A. Miyauchi, and T. Uemura (2006). Nanopillar sheets as a new type of cell culture dish: detailed study of HeLa cells cultured on nanopillar sheets, *J. Artif. Organs*, **9**, 90–96.
15. A. Miyauchi, K. Kuwabara, M. Hasegawa, and M. Ogino (2016). Large-area nanoimprint and application to cell cultivation, *Appl. Phys. A*, **122**, 265.
16. R. Takahashi, H. Sonoda, Y. Tabata, and A. Hisada (2010). Formation of hepatocyte spheroids with structural polarity and functional bile canaliculi using nanopillar sheets, *Tissue Eng. Part A*, **16**(6), 1983–1995.

17. K. Kuwabara, M. Ogino, T. Ando, and A. Miyauchi (2008). Enhancement of fluorescence intensity from an immunoassay chip using high-aspect-ratio nanopillars fabricated by nanoimprinting, *Appl. Phys. Lett.*, **93**(3), 033904.

18. K. Kuwabara, and A. Miyauchi (2008). High-aspect-ratio nanopillar structures fabricated by nanoimprinting with elongation phenomenon, *J. Vac. Sci. Technol. B*, **26**(2), 582–584.

19. S. Y. Chou, P. R. Kauss, W. Zhang, L. Guo, and L. Zhuang (1997). Sub-10 nm imprint lithography and applications, *J. Vac. Sci. Technol. B*, **15**, 2897–2904.

20. W. Wu, B. Cui, X. Sun, W. Zhang, L. Zhuang, L. Kong, and S. Y. Chou (1998). Large area high density quantized magnetic disks fabricated using nanoimprint lithography, *J. Vac. Sci. Technol. B*, **16**, 3825–3829.

21. M. Colbrun, S. C. Johnson, M. D. Stewart, S. Damie, T. C. Bailey, B. Choi, M. Wedlake, T. B. Michaelson, S. V. Sreenivasa, J. G. Ekerdt, and C. G. Willson (1999). Step and flash imprint lithography: a new approach to high-resolution patterning, *Proc. SPIE Int. Soc. Opt. Eng.*, **3676**(1), 379–389.

22. Q. Xia, J. J. Yang, W. Wu, X. Li, and R. S. Williams (2010). Self-aligned memristor cross-point arrays fabricated with one nanoimprint lithography step, *Nano Lett.*, **10**, 2909–2914.

23. S. Pi, P. Lin, and Q. Xia (2013). Cross point arrays of 8 nm × 8 nm memristive devices fabricated with nanoimprint lithography, *J. Vac. Sci. Technol. B*, **31**, 06FA02.

24. S. Y. Chou, P. R. Kraus, and P. J. Renstrom (1996). Nanoimprint lithography, *J. Vac. Sci. Technol. B*, **14**(6), 4129–4133.

25. T. K. Widden, D. K. Ferry, M. N. Kozicki, E. Kim, A. Kumar, J. Wilbur, and G. M. Whitesides (1996). *Nanotechnology*, **7**, 447–451.

26. H. Schift and A. Kristensen (2010). Nanoimprint lithography: patterning resists using molding. In *Handbook of Nanotechnology*, 271–312, Vol. ed. B. Bhushan, 3rd edition, Springer-Verlag, Berlin, Heidelberg.

27. M. Colburn, S. Johnson, M. Stewart, S. Damle, T. Bailey, B. Choi, M. Wedlake, T. Michaelson, S. V. Sreenivasan, J. Ekerdt, and C. G. Willson (1999). Step and flash imprint lithography: a new approach to high-resolution patterning, *Proc. SPIE*, Emerging Lithographic Technologies III, 379.

28. M. Colburn, I. Suez, B. J. Choi, M. Meissl, T. Bailey, S. V. Sreenivasan, J. G. Ekerdt, and C. G. Willson (2001). Characterization and modeling

of volumetric and mechanical properties for step and flash imprint lithography photopolymers, *J. Vac. Sci. Technol. B*, **19**(6), 2685.

29. M. Hatano, K. Kobayashi, H. Kashiwagi, H. Tokue, T. Kono, T. Nakasugi, E. H. Choi, and W. Jung (2016). NIL defect performance toward high volume mass production, *Proc. SPIE*. **9777**, Alternative Lithographic Technologies VIII, 97770B.
30. M. Colburn, T. Bailey, B. J. Choi, J. G. Ekerdt, and S. V. Sreenivasan (2001). Development and advantages of step-and-flash lithography, *Solid State Technol.*, **67**.
31. T. C. Bailey, D. J. Resnick, D. Mancini, K. J. Nordquist, W. J. Dauksher, E. Ainley, A. Talin, K. Gehoski, J. H. Baker, B. J. Choi, S. Johnson, M. Colburn, S. V. Sreenivasan, J. G. Ekerdt, and C. G. Willson (2002). Step and flash imprint lithography: an efficient nanoscale printing technology, *Microelectron. Eng.*, **61–62**, 461–467.
32. S.V. Sreenivasan, P. Schumaker, B. Mokaberi-Nezhad, J. Choi, J. Perez, V. Truskett, F. Xu, and X. Lu (2009). High-resolution defect inspection of step-and-flash imprint lithography for 32-nm half-pitch patterning, Presented at the SPIE Advanced Lithography Symposium, Conference 7271.
33. K. Selenidis, J. Maltabes, I. McMackin, J. Perez, W. Martin, D. J. Resnick, and S. V. Sreenivasan (2007). Defect reduction progress in step and flash imprint lithography, *Proc. SPIE*, **6730**, 67300F-1.
34. I. McMackin, J. Choi, P. Schumaker, V. Nguyen, F. Xu, E. Thompson, D. Babbs, S. V. Sreenivasan, M. Watts, and N. Schumaker (2004). Step and repeat UV nanoimprint lithography tools and processes, *Proc. SPIE*, **5374**, 222.
35. T. Higashiki, T. Nakasugi, and I. Yoneda (2011). Nanoimprint lithography for semiconductor devices and future patterning innovation, *Proc. SPIE*, **7970**.
36. Z. Ye, K. Luo, J. W. Irving, X. Lu, W. Zhang, B. Fletcher, W. Liu, M. Shafran, S. Lee, W. Longsine, V. Truskett, F. Xu, D. LaBrake, Douglas Resnick, and S. V. Sreenivasan (2013). Defect reduction for semiconductor memory applications using jet and flash imprint lithography, *Proc. SPIE*, **8680**, Alternative Lithographic Technologies V, 86800C.
37. W. Zhang, B. Fletcher, E. Thompson, W. Liu, T. Stachowiak, N. Khusnatdinov, J. W. Irving, W. Longsine, M. Traub, V. Truskett, D. LaBrake, and Z. Ye (2016). High throughput jet and flash imprint lithography for semiconductor memory applications, *Proc. SPIE*, **9777**, Alternative Lithographic Technologies VIII, 97770A.

38. B. J. Choi et al. (2005). Distortion and overlay performance of UV step and repeat imprint lithography, *Microelectron. Eng.*, **V78–79**, 633.

39. K. S. Selinidis, C. B. Brooks, G. F. Doyle, L. Brown, C. Jones, J. Imhof, D. L. LaBrake, D. J. Resnick, and S. V. Sreenivasan (2011). Mask replication using jet and flash imprint lithography, *J. Micro/Nanolith. MEMS MOEMS*, **10**(4), 043005 (Oct–Dec 2011).

40. Koji Ichimura, Kouji Yoshida, Saburo Harada, Takaharu Nagai, Masaaki Kurihara, and Naoya Hayashi (2016). Development of nanoimprint lithography templates toward high-volume manufacturing, *J. Micro/Nanolith. MEMS MOEMS*, **15**(2), 021006 (Apr–Jun 2016).

Index